Series / Number 07-051

STOCHASTIC PARAMETER REGRESSION MODELS

PAUL NEWBOLD
University of Illinois at Urbana-Champaign

THEODORE BOS
University of Alabama at Birmingham

Printed in the United States of America

For information address:

SAGE Publications, Inc.
2111 West Hillcrest Drive
Newbury Park, California 91320

SAGE Publications Ltd.
28 Banner Street
London EC1Y 8QE
England

SAGE Publications India Pvt. Ltd.
M-32 Market
Greater Kailash I
New Delhi 110 048 India

International Standard Book Number 0-8039-2425-9

Library of Congress Catalog Card No. 85-050149

THIRD PRINTING, 1990

When citing a professional paper, please use the proper form. Remember to cite the correct Sage University Paper series title and include the paper number. One of the following formats can be adapted (depending on the style manual used):

(1) IVERSEN, GUDMUND R. and NORPOTH, HELMUT (1976) "Analysis of Variance." Sage University Paper series on Quantitative Applications in the Social Sciences, 07-001. Beverly Hills and London: Sage Pubns.

OR

(2) Iversen, Gudmund R. and Norpoth, Helmut. 1976. *Analysis of Variance*. Sage University Paper series on Quantitative Applications in the Social Sciences, series no. 07-001. Beverly Hills and London: Sage Pubns.

CONTENTS

Series Editor's Introduction

Newbold and Bos have written an excellent introduction to stochastic parameter regression models. These models allow relationships to vary through time, rather than requiring them to be fixed as in most ordinary regression models, or forcing the analyst to specify and analyze the causes of the time-varying relationships. As such, this monograph will be most useful to social scientists who have a good working knowledge of standard regression models and wish to understand methods dealing with relationships that vary slowly over time but in which the exact causes of variation cannot be identified. This monograph, then, represents one of the more advanced and technically difficult works published to date in this series. We hope to continue to publish more introductory treatments but occasionally to introduce a more advanced and difficult topic, in order to improve the "state of the art" in quantitative social science.

Chapter 1 begins with a discussion of the ordinary linear regression model, then introduces stochastic parameter regression models (SPRMs) and univariate and vector autoregressive models. It provides an intuitive approach to the likelihood ratio and Lagrange multiplier tests. Chapter 2 discusses the Kalman filter, derives the exact log-likelihood of the SPRM, and presents estimation of the stochastic parameters for observed time periods as well as the prediction of their future values.

In Chapter 3, the authors discuss various testing devices for the SPRM and the model from which it was generalized, the random coefficient model (RCM). This chapter ends with an application of the tests in the estimation of an RCM and SPRM for the market model of finance theory. Chapter 4 covers a test of the hypothesis of efficiency in a capital market—the market for 3-month Treasury bills. The SPRM provides an interesting test of this hypothesis while also testing whether or not the equilibrium real rate of return is constant, thus demonstrating the advantages of SPRM.

The major audience for this monograph will no doubt be economists and econometricians, but it will doubtless also be of interest to psychometricians and to social scientists working in the growing field of political economy. It will also be useful for policy analysts working in government or private research organizations.

—*John L. Sullivan*
Series Co-Editor

STOCHASTIC PARAMETER REGRESSION MODELS

PAUL NEWBOLD
University of Illinois at Urbana-Champaign
THEODORE BOS
University of Alabama at Birmingham

1. INTRODUCTION AND PRELIMINARIES

Regression analysis is perhaps the most commonly employed tool of statistical model building in the social sciences. Suppose that we want to explain the behavior of some variable Y, which will be called a *dependent variable.* Subject matter theory might suggest that K *independent variables,* $X_1, X_2, \ldots, X_K$, influence the behavior of the dependent variable. Very often, in the social sciences, subject matter theory is imprecise as to the functional form of such relationships. Frequently, in such circumstances a *linear* model can provide a useful approximation, at least within the ranges of interest of the independent variables.

In order to model a relationship, data are required, and it will be assumed that T sets of observations $(y_1, x_{11}, x_{21}, \ldots, x_{K1})$, $(y_2, x_{12}, x_{22}, \ldots, x_{K2})$, $\ldots$, $(y_T, x_{1T}, x_{2T}, \ldots, x_{KT})$ are available on the dependent and independent variables. Data may be available across space, through time, or both. In much of our analysis, these T observations will be assumed to be taken at equally spaced points in time. In the *multiple linear regression model,* the data generation mechanism is taken to be of

$$y_t = \beta_1 x_{1t} + \beta_2 x_{2t} + \ldots + \beta_K x_{Kt} + e_t \qquad [1.1]$$

In (1.1), the β_i $(i = 1,2, \ldots, K)$ are fixed coefficients whose values can be estimated from the available data. The parameter β_i is interpreted as the expected increase in the dependent variable, following from a one unit increase in the ith independent variable, when the values of the other independent variables are held fixed. In equation (1.1), e_t is a stochastic *error term:* The presence of this random variable acknowledges the fact that, in practice, no relationship will hold precisely. Thus, e_t represents the difference between the observed value of the dependent variable and its expected value under the theoretical model. By setting $x_{1t} = 1$ (t=1,2, . . . ,T), an intercept term β_1 is introduced into the regression model.

At least as a first step, the multiple linear regression model is generally analyzed under the following set of standard assumptions.

(1) Either the x_{it} are fixed numbers, or, if they are random variables, they are independent of the error terms e_t. In the latter case analysis is conditioned on the observed values of the independent variables.
(2) The error terms e_t all have mean zero.
(3) The error terms e_t have a common, generally unknown, variance σ^2.
(4) The error terms e_t are not correlated with one another.
(5) There does not exist a set of fixed numbers $c_1, c_2, \ldots, c_K$ such that $c_1 x_{1t} + c_2 x_{2t} + \ldots + c_K x_{Kt} = 0$ for all t = 1,2, . . . ,T. In the absence of this assumption it is not possible to estimate the separate influences of the independent variables on the dependent variable.

Using matrix notation, the set of equations (1.1) can be written

$$y_t = (x_{1t}, x_{2t}, \ldots, x_{Kt}) \begin{pmatrix} \beta_1 \\ \beta_2 \\ \vdots \\ \beta_K \end{pmatrix} + e_t$$

$$= x_t' \beta + e_t$$

More compactly, we can write

$$\begin{pmatrix} y_1 \\ y_2 \\ \vdots \\ y_T \end{pmatrix} = \begin{bmatrix} x_{11} & x_{21} & \cdots & x_{K1} \\ x_{12} & x_{22} & \cdots & x_{K2} \\ \cdot & \cdot & & \cdot \\ x_{1T} & x_{2T} & & x_{KT} \end{bmatrix} \begin{pmatrix} \beta_1 \\ \beta_2 \\ \vdots \\ \beta_K \end{pmatrix} + \begin{pmatrix} e_1 \\ e_2 \\ \vdots \\ e_T \end{pmatrix}$$

or

$$y = X\beta + e$$

Now, if the five standard assumptions hold, it is known, by the *Gauss-Markov theorem,* that the least squares estimator of β,

$$b = (X'X)^{-1} X'y$$

is optimal in the sense that it is *best linear unbiased.* This implies that, for any fixed numbers $d_1, d_2, \ldots, d_K$, of all estimators of

$$d_1\beta_1 + d_2\beta_2 + \ldots + d_K\beta_K$$

that are linear in the y_t and unbiased, the estimator

$$d_1 b_1 + d_2 b_2 + \ldots + d_K b_K$$

has smallest variance.

An unbiased estimator of the common variance of the error terms e_t in (1.1) is obtained as

$$s^2 = \frac{\sum_{t=1}^{T} (y_t - b_1 x_{1t} - b_2 x_{2t} - \ldots - b_K x_{Kt})^2}{T - K}$$

$$= (y - Xb)'(y - Xb)/(T - K)$$

If, in addition to the other five assumptions, it is assumed that the error terms e_t have normal distributions, it can be shown that the random variables

$$t_i = \frac{b_i - \beta_i}{s\sqrt{a_{ii}}}$$

where a_{ii} is the ith diagonal element of the matrix $(X'X)^{-1}$, have Student's t distribution with T-K degrees of freedom. This result allows the derivation of confidence intervals for, or test of hypotheses on, the regression coefficients.

One application of regression models is to the *prediction* of future values of the dependent variables, given future values of the independent variables. Suppose that, standing at time T, we are looking forward l time periods, and want to forecast y_{T+l}. Assuming that the model (1.1) continues to hold in the future, we can write

$$y_{T+l} = \beta_1 x_{1,T+l} + \beta_2 x_{2,T+l} + \ldots + \beta_K x_{K,T+l} + e_{T+l} \qquad [1.2]$$

The optimal forecast of y_{T+l} is then obtained by substituting the least squares estimates b_j for the unkown parameters β_j in (1.2) and replacing e_{T+l} by its expected value of zero. The best linear unbiased predictor of y_{T+l} is, then,

$$b_1 x_{1,T+l} + b_2 x_{2,T+l} + \ldots + b_K x_{K,T+l}$$

If the error terms e_t are assumed to be normally distributed, then confidence intervals can be put around these point forecasts.

Now, the optimality of the least squares coefficient estimators and of the above forecasts depends on the validity of the assumptions made, as do the associated confidence intervals and tests of hypotheses. If these assumptions break down, least squares procedures can be seriously suboptimal, and inferential statements based on them badly misleading. It is, therefore, important to test, where possible, the assumptions. In particular, the assumptions that the error terms have the same variance and are not correlated with one another can be tested. Error terms with different variances are said to exhibit *heteroscedasticity.* When time-series data are analyzed, the assumption that the error terms are not correlated with one another is suspect. An alternative possibility is that errors in adjacent time periods will be somewhat similar, and therefore positively correlated with one another. This phenomenon, known as *autocorrelated errors,* is generally tested for through the *Durbin-Watson test,* the statistic for which is routinely produced by standard multiple regression computer programs.

Our objective here is to consider a different, but not unrelated, elaboration of the usual multiple linear regression model and its standard assumptions. It is generally assumed that the regression coefficients, β_i, are *fixed,* either across space or through time. We wish, in the context of our time-series model, to consider the possibility that the regression coefficients vary through time. Thus, at time t, the coefficient on the ith independent variable will be denoted β_{it}, to indicate that, at some other point in time, a different coefficient might be appropriate. Therefore, in place of equation (1.1) we write

$$y_t = \beta_{1t}x_{1t} + \beta_{2t}x_{2t} + \ldots + \beta_{Kt}x_{Kt} + e_t \qquad [1.3]$$

Essentially, by contemplating this elaboration of the usual multiple linear regression model, we are asserting that a unit change in one of the independent variables, all else equal, will not have a constant expected effect on the dependent variable at all points in time.

1.1 Regression Models with Stochastic Parameters

Like much quantitative methodology in the social sciences, regression analysis was taken over from the physical sciences, where it has long been widely and successfully employed. In both the physical and social sciences, subject matter theory is used to suggest a model describing the behavior of a variable or variables of interest. Data are then employed to estimate the parameters of the specified model. There is, however, a crucial distinction between much physical scientific and social scientific theory. In the physical sciences, theory often suggests the presence of *physical constants,* yet only rarely is this truly the case in the social sciences. For example, we know that if a heavy object is dropped to the ground today in Champaign, Illinois, it will accelerate at a rate of approximately 32 feet per second per second. Moreover, the rate of acceleration would be very nearly the same in Melbourne, Australia today, or in Champaign, Illinois in twenty years time. It is very difficult to think of any social scientific theory that strongly asserts the existence of such a physical constant. In economics, theories of the consumption function and of consumer demand are well developed. It is not, however, asserted as a result of such theories, that the marginal propensity to consume is the same in the United States as in Australia, or that the elasticity of demand for coconuts is the same in Champaign as in Melbourne. Neither, indeed, do such theories claim a constancy through time. Thus, there are no strong theoretical grounds for believing that the marginal propensity to consume in the United States, or the elasticity of demand for coconuts in Champaign, will reman fixed over the years.

In spite of these considerations, the multiple linear regression model with fixed coefficients (1.1) is frequently fitted to data by social scientists. As we have just suggested, one should be skeptical about the assumption, implicit in such analyses, about the constancy through time of the model parameters. One must doubt whether human behavior is as undeviating as such a formulation suggests. We do not intend to claim that the fixed parameters model is invariably inappropriate in social scientific research. A well-conceived statistical model should not be an attempt to describe *all* of the complex behavior in a system, but rather is an attempt to describe as efficiently as possible the *essence* of that system. Therefore, although it is difficult to believe in the existence of physical constants in the social sciences, it may nevertheless be the case that a regression model with fixed parameters is capable of yielding valuable insights into, and making good predictions about, social systems. However, it does seem prudent to ask whether a model in which the parameters are allowed to vary over time might provide a superior representation.

Once the assumption that the coefficients of a regression model are fixed is questioned, it is necessary to think about how these coefficients might change. One possibility is that, as a result of exogenous shocks to the system, there could be, over any time period, one or more abrupt changes in the values taken by the regression parameters. On occasion this kind of behavior might be suspected. However, the models studied here, which exhibit more stability, will often be more plausible alternatives to the fixed parameters specification.

Essentially, we will regard the parameters β_t *as random variables,* drawn through time from a common distribution. Models of the form (1.3) where the regression parameters are taken to be random variables are called *stochastic parameters regression models.* In general, we will not want to assume that the random variables β_t (t = 1,2, . . . ,T) are *independent* of one another. For example, the elasticity of demand for coconuts in Champaign, Illinois may be most sensibly thought of as changing over time. However, such changes are quite likely to be smooth, so that the value next year would be strongly correlated with this year's value. Thus, we would wish to consider the possibility that the stochastic parameters β_t are *autocorrelated.*

To summarize our discussion to this point, our objective is to study regression models of the form (1.3) where the parameters β_t are possibly autocorrelated random variables. Before proceeding to do so, however, we must pause to discuss means for modeling autocorrelation in random variables.

1.2 Models for a Single Time Series

In this monograph, interest is concentrated on the multiple regression model in which the regression parameters rather than being fixed may follow some stochastic process. However, before proceeding in Chapter 2 to a discussion of that model, it is necessary to introduce some results on the time-series models that will be employed in describing the evolution of the stochastic parameters. In this section we discuss models for a single time series; the following section briefly introduces the multivariate case. Although our discussion is centered on the issue of stochastic parameters, the results of this section apply equally to the modeling of an observed time series or of the autocorrelation structure of the error terms in a regression equation.

Let β_t be a random variable, representing observations through time on a single entity. In order to make progress in the study of such a process, some simplifying assumptions are necessary. An assumption that very often has been useful in the study of actual time-series data is of *stationarity*. The series β_t is said to be stationary if the following three conditions hold:

(1) The mean of β_t is the same for all time periods t.
(2) The variance of β_t is the same finite value for all time periods t.
(3) The correlation between the random variables β_t and β_s depends only on their distance apart in time; that is, on $|t-s|$.

The assumption of stationarity of a time series is by no means invariably appropriate. For example, on occassion it is preferable to assume that period to period changes, $\beta_t - \beta_{t-1}$, constitute a stationary process, rather than taking β_t to be stationary. However, in the present context, the stationarity assumption will generally be convenient, and we will proceed on that basis.

Because the process β_t is taken to be stationary, a common mean through time is assumed. Thus, if β denotes this common mean, we can write

$$E(\beta_t) = \beta$$

for all time periods t.

A particularly simple model that could describe the behavior of a stationary time series is

$$\beta_t - \beta = a_t \qquad [1.4]$$

where the random variables a_t have zero mean, fixed variance through time, and, for $t \neq s$, a_t is uncorrelated with a_s. Such a process a_t is often called *white noise*. If the time series β_t obeys the simple model (1.4) then the β_t have common mean β, common variance, and are not correlated with one another.

As we have already asserted, however, we want to study time series that are autocorrelated. Consider, for example, product sales. It might be conjectured that sales this year are quite strongly correlated with last year's sales level, rather less strongly correlated with sales two years ago, and so on—the strength of the correlation between two values of the time series decreasing as their distance apart in time increases. Such an autocorrelation structure has a good deal of intuitive appeal. A particularly simple model that gives rise to such a structure is the *first-order autoregressive process.* We will denote by ϕ the correlation between adjacent values in time of the series, so that

$$\text{Corr.}\ (\beta_t, \beta_{t-1}) = \phi$$

Then, suppose that the correlation between values two time periods apart is ϕ^2, the correlation between values three times periods apart is ϕ^3, and so on, so that

$$\text{Corr.}\ (\beta_t, \beta_{t-j}) = \phi^j \quad (j = 1,2,3, \ldots) \qquad [1.5]$$

Such a process satisfies our desideratum that the strength of the correlation between values of a series is inversely related to their distance apart in time. Now, it can be shown that if the stationary time series β_t possesses the autocorrelation structure (1.5), it must be generated by the model

$$(\beta_t - \beta) = \phi\ (\beta_{t-1} - \beta) + a_t \qquad [1.6]$$

where a_t is white noise, and the correlation coefficient ϕ is less than one in absolute value. The model (1.6) is called a *first-order autoregressive model,* and is frequently used in practice to represent autocorrelated time-series behavior.

If the stationary time series β_t obeys the first-order autoregressive model (1.6) then, as we have seen, the mean of β_t is β, and the autocorrelation structure is given by (1.5). It can further be shown that the variance of β_t is

$$\sigma_\beta^2 = \frac{\sigma_a^2}{1 - \phi^2} \qquad [1.7]$$

where σ_a^2 is the variance of the white noise process a_t.

Our discussion of the first-order autoregressive model derived from the designation of ϕ as the correlation between adjacent observations through time. Of course, it is required that this correlation not exceed one in absolute value. When ϕ is less than one in absolute value, the expression (1.7) for the variance of B_t is finite, and the assumptions of stationarity are satisfied. Now, it is possible to write down a model of the form (1.6) in which ϕ exceeds one in absolute value (though it cannot now be interpreted as a correlation coefficient). Such models are sometimes called "explosive," and are briefly discussed by Granger and Newbold (1977: 38). However, these models are not useful for representing real data in the social sciences. On the other hand, the "border line nonstationary" case, in which $\phi = 1$ in (1.6) is useful. This model is known as the "random walk," and is best thought of as describing a situation in which changes from one time period to the next are white noise, and therefore unpredictable in terms of past changes.

If a time series is autocorrelated, this factor can be exploited in the production of forecasts of the future on the basis of current and past observations. In this context, the first-order autoregressive model has an interesting feature. Suppose we are standing at time T, so that observations $\beta_T, \beta_{T-1}, \beta_{T-2}, \ldots$ are available, and that it is required to predict the value l time periods into the future, β_{T+1}. Setting t equal to T + l in equation (1.6) we can write

$$\begin{aligned}(\beta_{T+1} - \beta) &= \phi(\beta_{T+l-1} - \beta) + a_{T+l} \\ &= \phi^2(\beta_{T+l-2} - \beta) + a_{T+l} + \phi\, a_{T+l-1} \\ &= \phi^3(\beta_{T+l-3} - \beta) + a_{T+l} + \phi\, a_{T+l-1} + \phi^2\, a_{T+l-2}\end{aligned}$$

Continuing in this way, we find

$$(\beta_{T+1} - \beta) = \phi^{\ell}(\beta_T - \beta) + (a_{T+l} + \phi\, a_{T+l-1} + \ldots + \phi^{\ell-1}\, a_{T+l}) \qquad [1.8]$$

Now, future values a_{T+j} ($j \geq 1$) of the white noise process a_t will be unpredictable, so that it follows from equation 1.8 that the optimal forecast of β_{T+1}, given information available at time T, is

$$\beta + \phi^{l}(\beta_T - \beta)$$

We see, then, that forecasts of future values of the series depend only on the most recent observation β_T.

For many time series, better forecasts are obtained through using more than just the most recent observation. An extension of the model (1.6) is the second-order autoregressive model

$$(\beta_t - \beta) = \phi_1(\beta_{t-1} - \beta) + \phi_2(\beta_{t-2} - \beta) + a_t$$

for which it can be shown that optimal forecasts of all future values depend only on the two most recent observations, β_t and β_{T-1}. The first-order autoregressive model is, of course, the special case of the second-order model with ϕ_2 equal to zero.

More generally, for any positive integer p, we can consider the autoregressive model of order p (AR(p) model)

$$(\beta_t - \beta) = \phi_1(\beta_{t-1} - \beta) + \ldots + \phi_p(\beta_{t-p} - \beta) + a_t$$

for which optimal forecasts of all future values depend only on the p most recent observations.

In analyzing actual data, time-series analysts have found useful a class of models of which autoregressive models constitute a subset. Let β_t be a stationary process with mean β, and p and q be non-negative integers. Then, β_t is said to obey an autoregressive-moving average process of order (p, q) if it is generated by the model

$$\begin{aligned}(\beta_t - \beta) - \phi_1(\beta_{t-1} - \beta) - \ldots\ldots - \phi_p(\beta_{t-p} - \beta)& \\ = a_t - \theta_1 a_{t-1} - \ldots - \theta_q a_{t-q}&\end{aligned} \qquad [1.9]$$

where a_t is white noise, and the ϕ_i ($i = 1,2,\ldots,p$) and θ_j ($j = 1,2,\ldots,q$) are fixed parameters. Given observations $\beta_1, \beta_2, \ldots, \beta_T$ on a time series, the objective of time-series model building is to fit to these data an appropriate model from the general class. These models can then be projected forward to derive forecasts of future values. Model building for a single time series is discussed in a number of books, including Box and Jenkins (1970), Nelson (1973), and Granger and Newbold (1977).

In principle, autoregressive-moving average models can be used to represent stochastic parameter behavior. However, for practical purposes, it is generally adequate—and certainly computationally convenient—to represent such behavior by low-order pure autoregressive models.

1.3 Models for Multiple Time Series

In the previous section, we briefly introduced some models commonly employed to describe the evolution through time of a single series. Essentially, the models can be thought of as *filters* that transform an autocorrelated, and therefore predictable, series to a non-autocorrelated, white noise series. These white noise innovations may be viewed as those parts of the series that cannot be predicted from its immediate past. Similar considerations are pertinent when considering models for a set of related time series.

Suppose, now, that we have observations through time on a set of K possibly related time series. We will denote by

$$\beta_t' = (\beta_{1t}, \beta_{2t}, \ldots, \beta_{Kt})$$

the vector of observations on these variables at time t. As in Section 1.2, we will assume stationarity, which, in the present context, implies that the vector random variable β_t has a fixed mean vector and covariance matrix through time, and that, for every i and j, the covariance between β_{it} and $\beta_{j,t-l}$ depends only on *l*. We will denote by

$$\beta' = (\beta_1, \beta_2, \ldots, \beta_k)$$

the common mean of the random variables β_t' so that

$$E(\beta_t) = \beta$$

A natural generalization of the first-order autoregressive model in the scalar case is the *vector first-order autoregressive model*

$$\begin{pmatrix}\beta_{1t}\\ \beta_{2t}\\ \vdots\\ \beta_{Kt}\end{pmatrix}-\begin{pmatrix}\beta_{1}\\ \beta_{2}\\ \vdots\\ \beta_{K}\end{pmatrix}=\begin{bmatrix}\phi_{11} & \phi_{12} & \cdots & \phi_{1K}\\ \phi_{21} & \phi_{22} & \cdots & \phi_{2K}\\ \cdot & \cdot & & \cdot\\ \phi_{K1} & \phi_{K2} & \cdots & \phi_{KK}\end{bmatrix}\left[\begin{pmatrix}\beta_{1,t-1}\\ \beta_{2,t-1}\\ \vdots\\ \beta_{K,t-1}\end{pmatrix}-\begin{pmatrix}\beta_{1}\\ \beta_{2}\\ \vdots\\ \beta_{K}\end{pmatrix}\right]+\begin{pmatrix}a_{1t}\\ a_{2t}\\ \vdots\\ a_{Kt}\end{pmatrix},$$

which can be written more compactly as

$$\beta_t - \beta = \Phi(\beta_{t-1} - \beta) + a_t \qquad [1.10]$$

In equation (1.10), Φ is $K \times K$ matrix of fixed coefficients, and a_t is a K-dimensional white noise vector, with zero mean, fixed covariance matrix, denoted by Ω, such that each element of this vector at one time period is uncorrelated with each element at any other time period. Thus, with

$$a_t' = (a_{1t}, a_{2t}, \ldots, a_{Kt})$$

we have

$$E(a_t) = 0$$

$$E(a_t\, a_t') = \Omega$$

and

$$E(a_t\, a_{t-j}') = 0$$

for all j different from zero. Generalizing the result (1.7) of Section 1.2, if Ω_β denotes the covariance matrix of β_t, so that

$$\Omega_\beta = E[\,(\beta_t - \beta)\,(\beta_t - \beta)'\,]$$

it can be shown that

$$\Omega_\beta = \Phi\, \Omega_\beta\, \Phi' + \Omega$$

Suppose that, standing at time T, it is required to forecast the next set of values β_{T+1} of these time series. Then, extending the results of the previous section, it can be shown that, if the first-order vector autoregressive model is appropriate, the optimal forecast of β_{T+1} is

$$\beta + \Phi\,(\beta_T - \beta)$$

More generally, the optimal forecast of β_{T+l}, for any positive lead time l, is

$$\beta + \Phi^l\,(\beta_T - \beta)$$

Thus, the optimal predictor of any future element of the series depends only on the most recent set of observations. Notice, however, that forecasts of $\beta_{i,T+l}$ may depend on all $\beta_{i,T}$, so that each series may be useful in the forecasting of future values of any one series.

A special case of the model (1.10) arises when the autoregressive coefficient matrix Φ is diagonal, so that

$$\Phi = \begin{bmatrix} \phi_{11} & 0 & \dots & 0 \\ 0 & \phi_{22} & \dots & 0 \\ . & . & & . \\ 0 & 0 & \dots & \phi_{KK} \end{bmatrix}$$

In that case it follows from equation (1.10) that we can write

$$\beta_{it} - \beta_i = \phi_{ii}\,(\beta_{i,t-1} - \beta) + a_{it} \qquad (i = 1,2,\dots,K) \qquad [1.11]$$

so that each individual series follows a first-order autoregressive process, and is optimally predicted on the basis of just its own past values. The K autoregressive processes in (1.11) are not, however, uncorrelated with one another. This position would only arise if the a_{it} were also uncorrelated with one another, so that the matrix Ω was diagonal.

The vector first-order autoregressive model is a member of a more general class of models that have been used to represent multiple time series. Let β_t be a stationary process with mean β and p and q be any non-negative integers. Then, generalizing to the multiple time series case

(1.9), β_t is said to follow a vector autoregressive-moving average process of order (p,q) if it is generated by the model

$$(\beta_t - \beta) - \Phi_1 (\beta_{t-1} - \beta) - \ldots - \Phi_p (\beta_{t-p} - \beta) = a_t - \Theta_1 a_{t-1} - \ldots - \Theta_q a_{t-q}$$

where a_t is vector white noise, and the Φ_i (i = 1, . . . ,p) and Θ_j (j = 1, . . . ,q) are K × K matrices of fixed parameters. Given observations on the series β_t, the objective of time-series model building is to fit to these data an appropriate model from the general vector autoregressive-moving average class. Multiple time-series model building for such problems is discussed by Tiao and Box (1981), and Jenkins and Alavi (1981). However, our present concern is not with this problem. In the context of regression models with stochastic parameters, the β_t will be *unobserved*. In consequence, as a practical matter, it is extremely difficult, given available data, to decide on an appropriate model for the stochastic parameters. As a result, an analyst will generally proceed with just one specific model, the first-order autoregressive process being the most common choice.

1.4 Likelihood Ratio and Lagrange Multiplier Tests

Our objective here is to consider the possibility that the parameters of a regression model may be stochastic rather then fixed. We will see how the coefficients of a stochastic parameter model can be estimated. In addition we will want to *test hypotheses* about such models. For example, it is useful to have available a test of the null hypothesis that the fixed parameter model is appropriate against the alternative of stochastic parameters.

Perhaps the most generally applied hypothesis testing procedure in practice is the *likelihood ratio test*. Suppose that a statistical model involves a set of unknown parameters. A null hypothesis may specify a set of K equality constraints on these parameters, although they are free to take any values under the alternative hypothesis. Parameter estimates can be obtained through the method of *maximum likelihood*. Coefficient estimates are those values for which the *likelihood function*—that is, the joint probability density function of the observations, viewed as a function of the parameters—is a maximum. We denote by L_1 the maximum value of the likelihood function under the alternative hypothesis, and by L_0 the maximum value of the likelihood function when the parameters are forced to satisfy the K constraints imposed by the null hypothesis. The likelihood ratio test statistic is then

$$LR = 2 (\log L_1 - \log L_0)$$

It is known that, in large samples, this statistic has, under the null hypothesis, a chi-square distribution with K degrees of freedom. The null hypothesis is rejected for large values of the test statistic. It can be shown, under very general conditions that, for large samples, the likelihood ratio test has strong optimality properties.

An alternative test procedure, the *Lagrange Multiplier Test*, due to Rao (1948) and Silvey (1959), is available. To see the relationship between the two tests, suppose we have a statistical model with just a single unknown parameter θ, and that the null hypothesis specifies that this parameter takes the specific value θ_o. Having collected data, it is found that the maximum likelihood estimate of θ is θ_1. In Figure 1.1 we show a graph of the log likelihood function. This function has a maximum $\log L_1$ at $\theta = \theta_1$; its value at $\theta = \theta_0$ is $\log L_0$. The greater the difference between $\log L_1$ and $\log L_0$, the more suspicious would we be of the null hypothesis—that θ is equal to θ_0—and it is on this difference that the likelihood ratio test is based.

Now, at its maximum point, the slope of the log likelihood function is, of course, zero. Moreover, the further is the hypothesized value θ_0 from the maximum likelihood estimate θ_1, the higher in absolute value will be the slope of the log likelihood function at this hypothesized value. This suggests that a test of the null hypothesis can be based on the slope of the log likelihood function at θ_0. The Lagrange multiplier test is based on the derivative of the log likelihood function evaluated at the hypothesized value θ_0. It can be shown that, for large sample sizes, this derivative has, under the null hypothesis, a normal distribution with mean zero.

Details of the Lagrange multiplier test, and of its application to some econometric problems, are given by Breusch and Pagan (1980). The Lagrange multiplier test enjoys the same optimality properties as the likelihood ratio test in large samples. It does, however, also have a potentially very useful practical advantage. In order to compute the Lagrange multiplier test statistic it is only necessary to estimate the model under the null hypothesis. In order to carry out a likelihood ratio test, the model must also be estimated under the alternative hypothesis, which in some practical applications can greatly increase its computational cost.

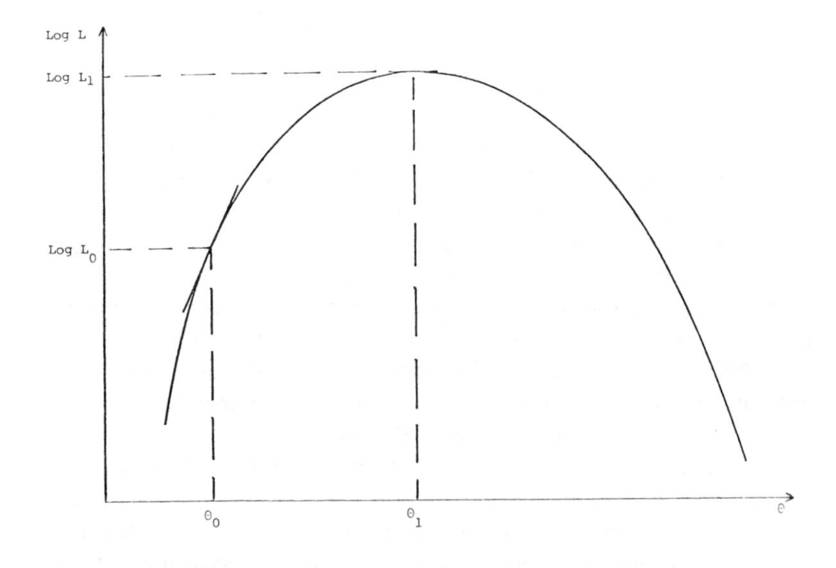

Figure 1.1: Illustration of Likelihood Ratio and Lagrange Multiplier Tests

An alternative test procedure, not employed in this monograph, is the *Wald test.* This shares optimality properties with the likelihood ratio and Lagrange multiplier tests, and requires estimation of the model only under the alternative hypothesis.

1.5 A Stochastic Parameters Regression Model

In the remainder of this monograph, we will examine the model

$$y_t = x_t'\beta_t + e_t \qquad [1.12]$$

where the vectors of parameters β_t are taken to be random variables. Although a number of alternative time-series models are available to describe the evolution through time of the stochastic parameters, we will restrict attention here to the first-order autoregressive model

$$\beta_t' - \beta = \Phi\,(\beta_{t-1} - \beta) + a_t \qquad [1.13]$$

On the basis of the types of data sets likely to be available in practice, it would be extremely difficult to distinguish among different members of the autoregressive-moving average class of models to represent stochastic parameter behavior, and we feel that the model (1.13) will generally prove adequate.

An important special case of this model arises when the matrix Φ of (1.13) is a matrix of zeros, so that

$$\beta_t - \beta = a_t$$

In that case, the parameters β_t can be viewed as random drawings from a common distribution with mean β, so that future values of the stochastic parameters are unpredictable from past values. This special case of the stochastic parameter model, known as the *random coefficients model* is frequently studied in practice. The random coefficients model will often be appropriate in the analysis of cross section data. Its estimation, using different procedures than those to be discussed in this monograph, is considered by Hildreth and Houck (1968). However, our aim is to concentrate here on the analysis of time-series data, where it is convenient to view the random coefficients model as a special case of the first-order autoregressive model that might, on occasion, arise in practice.

In Chapter 2 of this monograph we consider problems of estimation and prediction for the model (1.12) – (1.13), Chapter 3 introduces some useful tests of hypotheses. We conclude in Chapter 4 with a numerical example. The application of this model in the social sciences has been considered by a number of authors. Important references include Cooley and Prescott (1973), Engle and Watson (1981), Havenner and Swamy (1981), Pagan (1980), Rosenberg (1972, 1973), Swamy and Tinsley (1980), and Watson (1980).

2. ESTIMATION AND PREDICTION

As we discussed in the previous chapter, our objective is the study of a multiple regression model in which the parameters are not fixed, but rather are allowed to *evolve* through time, following a *stochastic* process. Suppose that we have T observations, $y_1, y_2, \ldots, y_T$ on a dependent variable, and, correspondingly, observations $x_{i1}, x_{i2}, \ldots, x_{iT}$ ($i = 1, 2, \ldots, K$) on each of K independent variables. We assume a linear relationship of the form

$$y_t = \beta_{1t} x_{1t} + \beta_{2t} x_{2t} + \ldots + \beta_{Kt} x_{Kt} + e_t \qquad [2.1]$$

where the β_{it} are time-varying parameters, and e_t is an error term assumed to have zero mean, fixed variance σ^2 through time, and to be non-autocorrelated. The model (2.1), then, differs from the usual multiple regression model under the standard assumptions only in that the parameters β_{it} are not restricted to be fixed over all time periods t. If we write the $1 \times K$ vectors

$$x_t' = (x_{1t}, x_{2t}, \ldots, x_{Kt}) \; ; \; \beta_t' = (\beta_{1t}, \beta_{2t}, \ldots, \beta_{Kt})$$

then equation (2.1) may be written more compactly as

$$y_t = x_t' \beta_t + e_t \qquad [2.2]$$

There are any number of ways in which the parameter vector β_t might evolve through time. As we discussed in Chapter 1, a first-order autoregressive model can often be useful for representing stochastic behavior of the kind in which we are interested here. We assume that the β_{it} are random variables, having fixed means β_i through time, so that the mean vector for β_t' is

$$\beta' = (\beta_1, \beta_2, \ldots, \beta_K)$$

If the parameters follow a first-order vector autoregressive process, then we can write

$$\beta_t - \beta = \Phi(\beta_{t-1} - \beta) + a_t \qquad [2.3]$$

where Φ is a $K \times K$ matrix of fixed parameters, and a_t is a $K \times 1$ vector of random errors, with zero mean and fixed covariance matrix Ω through time. The a_t are assumed to be non-autocorrelated, and also to be uncorrelated with the error terms in (2.2).

The pair of equations (2.2) and (2.3), together with the assumptions we have made about their error terms, fully define the stochastic parameter regression model we propose to analyze. As we have already noted, this model is a generalization of the usual fixed parameter regression model, the possibility of stochastic parameter evolution being allowed through (2.3). This model is a special case of a general class of models known as *state space* models. The use of such models is quite common in engineering, and in the last few years has attracted considerable attention in economics, finance, and the social sciences in general. In engineering terminology, equation (2.2) is called the ***measurement equation.*** It relates the observed values of the dependent variable to the independent variables, through the unobservable coefficients β_t. The ***transition equation*** (2.3) describes the evolution through time of these unobservable parameters.

To summarize, the model that we propose to analyze can be written as

$$y_t = \beta_{1t}\, x_{1t} + \beta_{2t}\, x_{2t} + \ldots + \beta_{Kt}\, x_{Kt} + e_t$$

where

$$\beta_t - \beta = \Phi(\beta_{t-1} - \beta) + a_t$$

In this specification, the innovation terms e_t and a_t are independent white noise processes. The independent variables x_{it} are either fixed numbers or random variables independent of e_t and a_t. In the special case in which the measurement equation contains an intercept term—as arises, for example, if we set $x_{1t} = 1$ $(t = 1,2,, \ldots ,T)$—an identification problem arises in the sense that it is impossible to distinguish the behavior of β_{1t} from that of the error term e_t, if the latter is also allowed to follow a first-order autoregressive process. In that case, the term e_t can be dropped from the model. However, we will proceed here with the more general formulation to incorporate the case of models with no intercept term.

Before proceeding with our analysis, it may be useful to look at the state space model in the more familiar multiple regression framework. We can write equation (2.2) as

$$y_t = x_t'\beta + x_t'(\beta_t - \beta) + e_t$$

or

$$y_t = x_t'\beta + u_t \quad [2.4]$$

where

$$u_t = x_t'(\beta_t - \beta) + e_t \quad [2.5]$$

Hence, (2.4) can be viewed as a multiple regression model with *fixed* parameters β, and a zero-mean error term u_t. However, if the β_t are not identically equal to their mean β—that is, if the β_t are stochastic—these error terms u_t will not all have the same variances, as it follows from (2.5) that the variance of u_t is

$$\mathrm{Var}(u_t) = x_t'\Omega_{\beta}x_t + \sigma^2$$

where Ω_{β} is the covariance matrix of β_t. Further, if the stochastic parameters β_t are autocorrelated—that is, if the matrix Φ of (2.3) is nonzero—then so will be the u_t of (2.4). Therefore, our stochastic parameter regression model can be regarded as a fixed parameter regression model with error terms that exhibit both heteroscedasticity and autocorrelation. The fixed parameters version of the model, with an error term exhibiting both heteroscedasticity and autocorrelated errors suggests the applicability of generalized least squares, which would be a relatively straightforward proposition if the parameters of the error covariance matrix were known.

This is not, however, the most convenient framework in which to analyze the model for purposes of parameter estimation and prediction. Rather, it is preferable to work with the state space form, and to employ a technique known as *Kalman filtering*. This provides a computationally efficient framework through which to attack the problems of estimation of the fixed parameters of the model, and prediction of future values. In this chapter we will look, using the Kalman filter, at three problems:

(1) The estimation of the fixed parameters of the model. Thus, we require estimates of the mean vector β of the stochastic regression coefficients, of the variance σ^2 of the error term in (2.2), of the covariance matrix Ω of the error term in (2.3), and also of the autoregressive parameter matrix, Φ, in that equation. We will

estimate the model parameters through maximum likelihood, employing the Kalman filter as a computationally efficient means of obtaining the likelihood function in terms of the fixed coefficients of the model.

(2) The prediction of future values of the stochastic parameters β_t, and of the dependent variables y_t for given future values of the independent variables.

(3) The estimation of the stochastic parameters β_t, $t = 1, 2, \ldots, T$, over the sample period.

We will consider these three problems in turn, but as a prelude we must introduce the Kalman filter. As will be seen, this provides a basis for the solution of our three problems.

2.1 The Kalman Filter

We proceed to discuss the stochastic parameter model (2.2) and (2.3) with the further assumption that the error terms in these two equations are normally distributed. The objective of the Kalman filter algorithm is then to provide a convenient way of computing the expected value and covariance matrix of the stochastic parameters β_t, *given information available at time t*. In order to attack this problem, it is necessary to introduce some further notation.

First, consider the distribution of β_t, given information up to time n, that is, given the observations $y_1, y_2, \ldots, y_n$. We will use $\beta(t|n)$ to denote the mean of this distribution and $P(t|n)$ to denote its $K \times K$ covariance matrix. Hence, in this notation,

$$\beta(t|n) = E[\beta_t|y_1, y_2, \ldots, y_n] \; ; \; P(t|n) = Var[\beta_t|y_1, y_2, \ldots, y_n]$$

We also need to consider the distribution of y_t, given information up to and including time $t - 1$. The mean and variance of this distribution will be written as

$$y(t|t-1) = E[y_t|y_1, y_2, \ldots, y_{t-1}] \; ; \; h_t = Var[y_t|y_1, y_2, \ldots, y_{t-1}]$$

We now proceed, in stages, to the derivation of the Kalman filter algorithm.

THE JOINT DISTRIBUTION OF β_t AND y_t, GIVEN INFORMATION AVAILABLE AT TIME $t - 1$

As a first step, it is necessary to find the joint distribution of β_t and y_t, given $y_1, y_2, \ldots, y_{t-1}$. Now, as a consequence of our normality assumption, it follows that this distribution is multivariate normal. Hence, the distribution is completely specified by its mean and covariance matrix.

Consider first, then, the mean of β_t. Taking expectations, conditional on the available information, in (2.3) yields

$$E(\beta_t | y_1,y_2, \ldots ,y_{t-1}) = \beta + \Phi E(\beta_{t-1} | y_1 y_2, \ldots ,y_{t-1}) - \Phi\beta + E(a_t | y_1,y_2, \ldots ,y_{t-1})$$

Now, because of our assumption of independence of the error terms in the measurement and transition equations, the final term on the right-hand side of this equation is zero, so that we can write

$$\beta(t|t-1) = \Phi\beta(t-1|t-1) + (I-\Phi)\beta \qquad (2.6]$$

where I is the $K \times K$ identity matrix.

It further follows from (2.3) that the covariance matrix of β_t, given information available at time $t - 1$, can be expressed as

$$\text{Var}(\beta_t | y_1,y_2, \ldots ,y_{t-1}) = \Phi \text{ Var}(\beta_{t-1} | y_1,y_2, \ldots ,y_{t-1})\Phi' + \Omega$$

where Ω is the covariance matrix of a_t. Hence, we can write

$$P(t|t-1) = \Phi P(t-1|at-1)\Phi' + \Omega \qquad [2.7]$$

The mean of the distribution of y_t, given information up to time $t-1$ is obtained by taking conditional expectations in (2.2), so that

$$E(y_t | y_1,y_2, \ldots ,y_{t-1}) = x_t' E(\beta_t | y_1,y_2, \ldots ,y_{t-1}) + E(e_t | y_1,y_2, \ldots ,y_{t-1})$$

Given our assumptions about the error term in (2.2), the final term on the right-hand side of this equation is zero, so that we can write

$$y(t|t-1) = x_t'\beta(t|t-1) \qquad [2.8]$$

Again using (2.2), the variance of y_t, given information available at time t – 1, is

$$\mathrm{Var}(y_t | y_1, y_2, \ldots, y_{t-1}) = x_t' \, \mathrm{Var}(\beta_t | y_1, y_2, \ldots, y_{t-1}) \, x_t + \sigma^2$$

where σ^2 is the variance of e_t. Therefore, we can write

$$h_t = x_t' P(t|t-1) x_t + \sigma^2 \qquad [2.9]$$

Finally, to complete the specification of this joint conditional distribution, we need the covariance between $\boldsymbol{\beta}_t$ and y_t, given information available at time t – 1. It follows from (2.2) that this can be expressed as

$$\begin{aligned} \mathrm{Cov}(\beta_t, y_t | y_1, y_2, \ldots, y_{t-1}) &= \mathrm{Var}(\beta_t | y_1, y_2, \ldots, y_{t-1}) \, x_t \qquad [2.10] \\ &= P(t|t-1) \, x_t \end{aligned}$$

THE DISTRIBUTION OF β_t,
GIVEN INFORMATION AVAILABLE AT TIME t

Next, we require the distribution of the stochastic parameters $\boldsymbol{\beta}_t$, given information available at time t—that is, given $y_1, y_2, \ldots, y_t$. This can be obtained from our previous results by noticing that what is needed is the distribution of $\boldsymbol{\beta}_t$, given $y_1, y_2, \ldots, y_{t-1}$, conditioned on the distribution of y_t given $y_1, y_2, \ldots, y_{t-1}$.

Using well-known properties of the multivariate normal distribution, it then follows that the conditional distribution we require is normal, with mean

$$\begin{aligned} \beta(t|t) = \Phi \, \beta(t-1|t-1) &+ (I-\Phi)\beta \\ &+ P(t|t-1) x_t h_t^{-1} [y_t - x_t' \beta(t|t-1)] \qquad [2.11] \end{aligned}$$

and covariance matrix

$$P(t|t) = P(t|t-1) - P(t|t-1) x_t h_t^{-1} x_t' P(t|t-1) \qquad [2.12]$$

These results are established in the appendix to this chapter.

THE KALMAN ALGORITHM

We are now in a position to collect our results to form a very convenient algorithm for computing the means and variances of these conditional distributions. Specifically, we require the set of equations (2.6), (2.7), (2.9), (2.11), and (2.12). Setting these together, we have the following:

$$\left.\begin{aligned}
&\text{(i)}\quad \beta(t\,|\,t-1) = \Phi\,\beta(t-1\,|\,t-1) + (I-\Phi)\,\beta \\
&\text{(ii)}\quad P(t\,|\,t-1) = \Phi\,P(t-1\,|\,t-1)\,\Phi' + \Omega \\
&\text{(iii)}\qquad h_t = x_t'\,P(t\,|\,t-1)\,x_t + \sigma^2 \\
&\text{(iv)}\qquad \beta(t\,|\,t) = \Phi\,\beta(t-1\,|\,t-1) + (I-\Phi)\,\beta \\
&\qquad\qquad + P(t\,|\,t-1)\,x_t h_t^{-1}\,[y_t - x_t'\beta(t\,|\,t-1)] \\
&\text{(v)}\qquad P(t\,|\,t) = P(t\,|\,t-1) - P(t\,|\,t-1)\,x_t h_t^{-1}\,x_t'\,P(t\,|\,t-1)
\end{aligned}\right\} \quad [2.13]$$

The set of equations (2.13) constitutes a special case of the Kalman filter algorithm, derived for a more general class of problems by Kalman (1960) and Kalman and Bucy (1961).

These equations allow the ready computation of the conditional means and variances of the stochastic parameters β_t, *for any given values* of the fixed parameters β, Φ, Ω and σ^2 of the state space model. The computations are started off by picking initial values for the vector $\beta(0|0)$ and $P(0|0)$. In many applications these initial values are chosen rather arbitrarily. However, as the computations proceed, the influence of this choice typically has a negligible impact on the final results. Given these starting values, we can, upon setting $t = 1$ in equations (2.13), use (i)-(v) in turn to find $\beta(1|0)$, $P(1|0)$, h_1, $\beta(1|1)$, and $P(1|1)$. Next, using these results, we can compute, by setting $t = 2$ in (2.13), $\beta(2|1)$, $P(2|1)$, h_2, $\beta(2|2)$, and $P(2|2)$. We can proceed iteratively in this fashion, setting in turn $t = 1, 2, \ldots, T$ in equations (2.13). The quantities required to perform the calculations at any one state will simply be the results found at the previous stage. Of course, these computations are arithmetically tedious. However, their repetitive nature makes them ideally suited for programming on an electronic computer, and the resulting algorithm is extremely efficient.

STARTING VALUES

In order to employ the Kalman filter algorithm (2.13), we need to specify the initial values $\beta(0|0)$ and $P(0|0)$. These are simply the mean and covariance matrix of the parameter vector β_0, given no observations. We know from the specification (2.3) that the stochastic parameters follow a vector first-order autoregressive process. Thus, the quantities we require are simply the mean and variance of such a process. From our discussion of the previous chapter, we know that the mean is simply

$$\beta(0|0) = \beta$$

and that the covariance matrix can be obtained by solving the set of equations

$$P(0|0) = \Phi P(0|0)\Phi' + \Omega$$

These starting values can be substituted into (2.13) to initialize the computations. These equations are then employed to compute recursively the means and variances of the dependent variables and the stochastic parameters in any time period, given information available in the previous period and the fixed coefficients of the model.

As we shall see, this algorithm provides us with a basis for the maximum likelihood estimation of the fixed coefficients, for the prediction of future values of the stochastic parameters and the dependent variables, and for the estimation of the stochastic parameters over the sample period.

2.2 Estimation of the Fixed Coefficients of the Model

Our state space model (2.2) and (2.3) involves a number of fixed parameters. These are the mean β of the stochastic parameters, the autoregressive matrix Φ of the process followed by these parameters, the error variance σ^2 of the measurement equation, and the covariance matrix Ω of the error terms in the transition equation. Given observations running over T time periods, we want to obtain estimates of these parameters. In this section we consider their estimation by the method of maximum likelihood.

We need, then, to derive the likelihood function—that is, the joint distribution of $y_1, y_2, \ldots, y_T$, as a function of the fixed parameters of the model. In the context of our model, Schweppe (1965) and Gupta and Mehra (1974) have shown how the Kalman filter algorithm (2.13) can be exploited for this purpose. Now, the joint probability density function of $y_1, y_2, \ldots, y_T$ can be expressed as the product of the conditional densities of y_t, given $y_1, y_2, \ldots, y_{t-1}$; that is, we can write

$$p(y_1, y_2, \ldots, y_T) = p(y_1)\, p(y_1)\, p(y_2 \mid y_1)\, p(y_3 \mid y_1, y_2) \ldots p(y_T \mid y_1, y_2, \ldots, y_{T-1})$$

Furthermore, we know that the distribution of y_t, given $y_1, y_2, \ldots, y_{t-1}$ is normal with mean given by (2.8) as $x'_t\beta(t|t-1)$, and variance h_t, given by (2.9). It therefore follows that the conditional density function is

$$P(y_t \mid y_1, y_2, \ldots, y_{t-1}) = (2\pi h_t)^{-\frac{1}{2}} \exp.[-(y_t - x'_t \beta(t \mid t-1))^2 / 2h_t]$$

It now follows that the likelihood function, or joint density function of the observations, is given by

$$L = (2\pi)^{-T/2} \prod_{t=1}^{T} h_t^{-\frac{1}{2}} \exp.\left[- \sum_{t=1}^{T} (y_t - x'_t \beta(t \mid t-1)^2 / 2h_t \right]$$

Taking logarithms then yields the log likelihood function, which is easier to work with computationally, as

$$\log L = -\frac{T}{2} \log(2\pi) - \frac{1}{2} \sum_{t=1}^{T} \log h_t - \frac{1}{2} \sum_{t=1}^{T} (y_t - x'_t \beta(t \mid t-1))^2 h_t^{-1} \qquad [2.14]$$

Equation (2.14), then, provides the function that must be maximized in order to obtain maximum likelihood estimates of the fixed coefficients of our state space model. Notice that, in addition to the observations y_t and x_t, this expression involves only the conditional expectation

$\beta(t|t-1)$ of the stochastic parameters, and the conditional variance h_t of the dependent variables. Both of these quantities may be conveniently evaluated as functions of the fixed parameters of our model through the Kalman algorithm (2.13). The Kalman filter, then, provides us with a very convenient and computationally efficient basis for deriving the likelihood at any set of values for the fixed parameters.

In order to derive maximum likelihood estimates of these parameters, we require the values for which the log likelihood function (2.14) is a maximum, or equivalently we need to minimize the function

$$\sum_{t=1}^{T} [\log h_t + (y_t - x_t' \beta(t|t-1))^2 h_t^{-1}]$$

Now, it is not possible to solve this function minimization problem analytically, as h_t and $\beta(t|t-1)$ will be rather complicated functions of the fixed parameters of the state space model. Rather, numerical function optimization algorithms must be employed. This, however, presents no great practical difficulty, as many computer subroutines have been written for such a purpose, and these algorithms are widely available. All that is required to employ such programs is a definition of the function to be optimized, which, as we have seen, is easily programmed through the Kalman filter algorithm. Numerical function minimization algorithms, such as the Newton-Raphson method require the derivatives of the function to be minimized. In fact, as discussed by Pagan (1980) and Bos (1982), it is possible to obtain these quantities analytically. The details of the derivation are, however, rather tedious, and will not be reproduced here. An alternative approach is to evaluate the derivatives numerically.

The numerical maximization of the log likelihood function yields point estimates of the fixed parameters of our model. Generally, it is also of value to calculate the standard errors associated with such estimators. As usual with maximum likelihood estimation, this can be achieved through the information matrix. Let θ denote the vector of all fixed coefficients, and L the likelihood function; then the information matrix $I(\theta)$ is defined in terms of the second partial derivatives of log L with respect to θ as

$$I(\theta) = -E \frac{\partial^2 \text{Log L}}{\partial\theta\, \partial\theta'}$$

where E denotes expectation. Then, an expression that is approximately valid in large sample sizes for the covariance matrix $V(\theta)$ of the maximum likelihood estimators of θ, is provided by

$$V(\theta) = I(\theta)^{-1}$$

In practice, it is necessary to substitute the maximum likelihood estimates for the unknown coefficients θ in this expression.

Pagan (1980) provides a computationally straightforward method for deriving the information matrix of a state space model. The details, however, are rather involved, and need not detain us here. An alternative is to evaluate numerically the second derivatives of the log likelihood function at its maximum point.

The speed with which modern electronic computers can carry out routine arithmetic computations allows us to analyze, without great difficulty, models such as the stochastic parameter model. Without the aid of computers, the computations discussed here would be prohibitively difficult. Our objective has been to outline the rationale behind procedures in current use for the estimation of the coefficients of the stochastic parameters regression model. It is important that the computational algorithms employed be as efficient as possible, and to this end we employ the Kalman filter as an aid in developing the likelihood function.

2.3 Prediction of Future Values

An important issue that frequently arises in empirical research concerns the *prediction* of future values. Let us assume that we have observed the independent variables $x_{1t}, x_{2t}, \ldots, x_{Kt}$, and the corresponding values y_t of the dependent variable, over the time periods $t = 1, 2, \ldots, T$. We will further assume that these data are generated by the stochastic parameter model (2.2) and (2.3), and that this model continues to hold true in the future. Our aim in this section is to develop predictions of the stochastic parameters and of the dependent variables over future time periods. These latter predictions will be *conditional forecasts;* that is, we derive predictions of the values that the dependent variable will take, given specific future values of the independent variables. One possible application of these procedures provides a quick way to generate regional economic forecasts. For example, suppose we want to predict future unemployment levels in the state of Illinois. The major determinant of future trends in Illinois is likely to be the overall national outlook for unemployment. We might, therefore, fit, to historical data, a regression equation in which the dependent variable is the unemployment rate in Illinois. A particularly simple model may have as the sole independent variable the rate of unemployment in the United States. Now, it would be unrealistic to assume at the outset a constant fixed relationship through time between the rates of unemployment in the state of Illinois and the nation as a whole. Accordingly, we could fit a regression model with stochastic parameters. Once this has been done, we will be in a position to derive conditional forecasts of the Illinois unemployment rate. As inputs to the forecast-generating mechanism, we can use one of the many widely available sets of forecasts of the national unemployment rate. Forecasts of future levels of unemployment in the state will then be conditioned on the national forecasts.

Given information available at time T, we want to predict the position, l time periods into the future, at time T + l. Let us assume, temporarily, that the values of the fixed coefficients of the model (2.2) and (2.3) are known. Thus, we take as given the parameters β, Φ, σ^2, and Ω.

To begin, let us consider the problem of predicting future values of the stochastic parameters. Now, our best forecast of β_{T+l} is the conditional expectation of that parameter, given information available at time T. Thus, in the notation established earlier, we need to find

$$\beta(T+l|T) = E[\beta_{T+l}|y_1, y_2, \ldots, y_T]$$

Such *point forecasts* are certainly valuable. However, it is important, though unfortunately by no means standard practice, to assess the amount of uncertainty associated with these forecasts. Accordingly, we can also compute the covariance matrix

$$P[T + l|T) = \mathrm{Var}[\beta_{T+l}|y_1, y_2, \ldots, y_T]$$

of the future stochastic parameters, given the available data. The diagonal elements of this matrix are then the variances of the forecast errors for the individual elements of the stochastic parameter vector. Assuming a normal distribution, the corresponding standard deviations can then be employed to compute *interval forecasts.*

The expected value of β_{T+l}, given information available at time T, is obtained by setting t equal to T+l in equation (2.3), and taking conditional expectations. We then find

$$E(\beta_{T+l}|y_1,y_2,...,y_T) = \beta + \Phi\, E(\beta_{T+1-l}|y_1,y_2,...,y_T) - \Phi\,\beta \\ + E(a_{T+l}|y_1,y_2,...,y_T)$$

Because the last term on the right-hand side of this equation is zero, we can write this expression as

$$\beta(T+l|T) = \Phi\beta(T+l-1|T) + (I-\Phi)\beta \qquad [2.15]$$

Equation (2.15) provides the basis for the generation of point forecasts of future values of the stochastic parameters. The forecasts are obtained one at a time, by setting in turn $l = 1, 2, 3, \ldots$ in (2.15). The point forecast for β_{T+1} is, therefore,

$$\beta(T+1|T) = \Phi\,\beta(T|T) + (I-\Phi)\beta$$

The conditional expectation $\beta(T|T)$ is the expected value of the stochastic parameters at the final sampling period, given all of the information in the sample. This value will have been found from the Kalman filter equations (2.13), which therefore provide the necessasry inputs to the forecast computations. Next, substituting $l = 2$ in (2.15), $\beta(T+1|T)$ is used to derive the point forecast of β_{T+2}. Continuing in this manner, point forecasts of the stochastic parameters as far ahead as desired can be calculated.

The covariance matrix of β_{T+l}, given information available at time T, can also be derived on setting t equal to T + l in (2.3). We then have

$$\mathrm{Var}(\beta_{T+l}|y_1,y_2,...,y_T) = \Phi\,\mathrm{Var}(\beta_{T+l-1}|y_1,y_2,...,y_T)\Phi' + \Omega$$

Therefore, we can write

$$P(T+l|T) = \Phi P(T+l-1|T)\Phi' + \Omega \qquad [2.16]$$

Equation (2.16) is used to compute the covariance matrices of the forecast errors for future values of the stochastic parameters, by setting in turn $l = 1, 2, 3, \ldots$ The calculations are initialized by employing P(T|T), obtained from the Kalman filter (2.13).

The forecasts of future values of the stochastic parameters are used in the derivation of forecasts of the dependent variables. We now require predictions of y_{T+l}, given all the information available at time T and also future values of the independent variables. As a point forecast of y_{T+l}, the optimal choice is again the conditional expectation of this random variable given all of the sample information. Therefore, we require

$$y(T+l|T) = E[y_{T+l}|y_1, y_2, \ldots, y_T]$$

Once again, it is also useful to have a measure of uncertainty associated with these point forecasts. Therefore, we require the variance of y_{T+l} given information available at time T. This quantity will be written

$$h(T+l|T) = \mathrm{Var}[y_{T+l}|y_1, y_2, \ldots, y_T]$$

Assuming a normal distribution, it is then possible to compute interval forecasts. For example, a 95% interval forecast for y_{T+l} is given by

$$y(T+l|T) \pm 1.96\ \sqrt{h(T+l|T)}$$

The mean of the conditional distribution of y_{T+l}, given information available at time T, is obtained by setting t equal to $T+l$ in equation (2.2), and taking conditional expectations, yielding

$$E(y_{T+l}|y_1, y_2, \ldots, y_T) = x'_{T+l}\ E(\beta_{T+l}|\ y_1, y_2, \ldots y_T)$$
$$+ E(e_{T+l}|y_1, y_2, \ldots, y_T)$$

Because the second term on the right-hand side of this expression is zero, we have

$$y(T+l|T) = x'_{T+l}\ \beta(T+l|T) \qquad [2.17]$$

Equation (2.17) expresses the point forecasts of future values of the dependent variable in terms of those of the stochastic parameters, and the given future values of the independent variables.

Finally, again using equation (2.2), we obtain the forecast error variances as

$$\mathrm{Var}(y_{T+l}|y_1, y_2, \ldots, y_T) = x'_{T+l}\ \mathrm{Var}(\beta_{T+l}|y_1, y_2, \ldots, y_T)\ x_{T+l} + \sigma^2$$

where σ^2 is the variance of the error term in (2.2). Therefore, we find

$$h(T+l|T) = x'_{T+l} P(T+l|T)x_{T+l} + \sigma^2 \qquad [2.18]$$

Hence, equation (2.18) allows us to compute interval forecasts for future values of the dependent variables.

Now, for any given values of the fixed coefficients β, Φ, σ^2, and Ω of the stochastic parameters regression model, it is, as we have just seen, a relatively straightforward matter to employ the Kalman filter to obtain point and interval forecasts of future values of the stochastic parameters and of the dependent variable. However, in practice it is almost invariably the case that these fixed coefficients will be unknown. They can be estimated from available data using the maximum likelihood procedure discussed in Section 2.2. These maximum likelihood estimates are then substituted for the corresponding unknown true coefficient values to allow the computation of the forecasts. A consequence of having to employ coefficient estimates in place of the true values is that expressions (2.16) and (2.18) are likely to understate somewhat the uncertainty associated with the point forecasts. The resulting interval forecasts will therefore tend to be rather too narrow. However, if coefficient estimation is based on a moderately large number of sample observations, this problem should not be too serious.

2.4 Estimation of the Sample Period Stochastic Parameters

If an analyst believes that a stochastic parameter regression model might provide a good description of a data set, then it is natural to require estimates that trace, as well as possible, the progression through time over the sample period of the stochastic parameters. Therefore, in terms of equation (2.2), we require estimates of the parameters β_t, over the period $t = 1, 2, \ldots, T$. These estimates should optimally be based on *all* the available sample information, so that the optimal estimate of β_t is its conditional expectation given the entire sample data, that is

$$\beta(t|T) = E[\beta_t|y_1,y_2, \ldots ,y_T]$$

In order to find the standard errors associated with these estimators, so that interval estimates might be derived, we also require the conditional covariance matrices

$$P(t|T) = Var[\beta_t|y_1,y_2, \ldots ,y_T]$$

Now, the Kalman filter algorithm (2.13) does not provide the estimates we require. Rather, that algorithm generates, as functions of the fixed coefficients of the model, the mean and variance of the stochastic parameters β_t, *given only information available up to time t.* Therefore, the only estimates we have up to this point that utilize all the available sample information are $\beta(T|T)$ and $P(T|T)$, the mean and covariance matrix of the stochastic parameter vector at the last point in the sampling period.

Our problem can be solved through the use of the *fixed interval smoothing algorithm.* The derivation of this result is rather involved, so that we will here state it without proof. Formal proofs are given by Sage and Melsa (1971), Anderson and Moore (1979), and Ansley and Kohn (1982), the last of these being the most straightforward. The required conditional means and covariance matrices of the stochastic parameters are obtained from

$$\begin{aligned} \beta(t \mid T) &= \beta(t \mid t) + A_t\,[\beta(t+1 \mid T) - \beta(t+1 \mid t)] \\ P(t \mid T) &= P(t \mid t) - A_t\,[P(t+1 \mid t) - P(t+1 \mid T)]\,A_t' \end{aligned} \qquad [2.19]$$

where

$$A_t = P(t|t)\,\Phi'\,P(t+1|t)^{-1}$$

The calculations are begun by setting $t = T - 1$ in (2.19), thus obtaining the mean and covariance matrix of the conditional distribution of β_{T-1}. Notice that this requires only information generated by the Kalman filter algorithm. The equations (2.19) can then be employed by setting in turn $t = T - 2, T - 3, T - 4, \ldots$ to obtain point estimates $\beta(t|T)$ of all the stochastic parameters over the sample period, together with the associated covariance matrices $P(t|T)$.

In principle, to carry out the calculations of the fixed interval smoothing algorithm equations (2.19), it is necessary to know the values of the fixed coefficients β, Φ, Ω, σ^2 of the model. However, in practice we will have to replace these unknown true values by their maximum likelihood estimates.

2.5 Some Extensions

In this chapter we have considered in detail problems of estimation and prediction in a multiple regression model with stochastic parameters. The formulation (2.2) and (2.3), that we have adopted for purposes of exposition, requires that all of the parameters be stochastic, obeying a vector first-order autoregressive process. In this section, we briefly discuss two extensions of this model. First, we show that the case in which the parameters on some of the independent variables are fixed through time—while the other parameters are stochastic—is easily handled within the framework already set out. Second, we consider the possibility that the stochastic parameters might obey some process other than the vector first-order autoregressive model.

A mixture of fixed and stochastic parameters. On occasions subject matter theory or other considerations might suggest that for some independent variables it is appropriate to treat the regression parameters as fixed, although the parameters corresponding to the remaining independent variables may be stochastic. Suppose that a regression involves K independent variables, $x_1, x_2, \ldots, x_{K_1}, x_{K_{1+1}}, \ldots x_K$ where K_1 is some integer less than K, and only the parameters corresponding to the first K_1 of these variables are to be regarded as stochastic. The regression model is then

$$y_t = \beta_{1t}x_{1t} + \ldots + \beta_{K_1t}x_{K_1t} + \beta_{K_1+1}x_{K_1+1,t} + \ldots + \beta_K x_{Kt} + e_t$$

where $\beta_{K_1+1}, \ldots, \beta_{K_1}$ are the fixed parameters. In vector notation, we can write this equation as

$$y_t = x_t^{(1)'} \beta_t^{(1)} + x_t^{(2)'} \beta^{(2)} + e_t \qquad [2.20]$$

where

$$x_t^{(1)'} = (x_{1t}, x_{2t}, \ldots, x_{K_1t}) \;;\; \beta_t^{(1)'} = (\beta_{1t}, \beta_{2t}, \ldots, \beta_{K_1t})$$

and

$$x_t^{(2)'} = (x_{K_1+1,t}, \ldots, x_{Kt}) \;;\; \beta^{(2)'} = (\beta_{K_1+1}, \ldots, \beta_K)$$

Assuming that the stochastic parameters obey a vector first-order autoregressive model, with mean

$$\beta^{(1)'} = (\beta_1, \beta_2, \ldots, \beta_{K_1})$$

the formulation of the model is completed by writing

$$\beta_t^{(1)} - \beta^{(1)} = \Phi\left(\beta_{t-1}^{(1)} - \beta^{(1)}\right) + a_t$$

where Φ now is a $K_1 \times K_1$ matrix of autoregressive coefficients.

Now, equation (2.20) can be written as

$$z_t = x_t^{(1)'} \beta_t^{(1)} + e_t$$

where

$$z_t = y_t - x_t^{(2)'} \beta^{(2)}$$

It follows, then, that our model is in precisely the same form as (2.2) – (2.3), with z_t in place of y_t. Hence, the Kalman filter can be used precisely as before to derive the likelihood function, which is then maximized to obtain estimates of the fixed coefficients of the model. Similarly, the procedures discussed earlier for the prediction of future values, and for the estimation of the stochastic parameters over the sample period, also may be applied here. We see, therefore, that the model involving a mixture of fixed and stochastic parameters may be analyzed in a straightforward manner using the techniques discussed in the earlier sections of this chapter.

Stochastic parameters generated by autoregressive-moving average models. When stochastic variation in regression parameters is allowed, the first-order autoregressive model is just one of many processes that might be employed to represent this variation. As we saw in the previous chapter, a more general linear model is the autoregressive-moving average model

$$\begin{aligned}\beta_t - \beta = {} & \Phi_1(\beta_{t-1} - \beta) + \ldots + \Phi_p(\beta_{t-p} - \beta) \\ & + a_t - \Theta_1 a_{t-1} - \ldots - \Theta_q a_{t-q}\end{aligned} \qquad [2.21]$$

where the Φ_i and Θ_j are $K \times K$ matrices of fixed coefficients, and a_t is a $K \times 1$ vector of random errors, as in our earlier formulation. The quantities p and q are integers, determining, respectively, the order of the autoregressive and moving average parts of the model. Of course, in the special case where p is one and q is zero, we have the first-order autoregressive model.

In principle, as discussed for example by Swamy and Tinsley (1980), the procedures of this chapter can be applied to the more general case, in which the stochastic parameters are generated by an autoregressive-moving average model. However, in practice, such an approach involves two difficulties. First, as the number of fixed coefficients increases, their estimation through the numerical maximization of the likelihood function becomes very expensive. Second, it is likely to prove very difficult, on the basis of only moderately long data series, to choose an appropriate model from the general autoregressive-moving average class—that is, to select the values of p and q in (2.21). This problem can be challenging enough when one is attempting to fit such models to observed time series. It is considerably harder in the present context, where the β_t are not directly observable. Our own view is that, for the great majority of problems met in practice, a regression model with stochastic parameters following a first-order autoregressive process should provide an adequate representation of the available data, though on occasion one or more additional autoregressive terms may be required.

2.6 Appendix

In this appendix, we derive the results (2.11) and (2.12), which form the basis for the Kalman filter algorithm. To do so, we introduce a well-known result on the multivariate normal distribution. Let z_1 and z_2 be a pair of vectors of random variables, whose joint distribution is multivariate normal. These random variables will be taken to have means and covariance matrices μ_i and Σ_{ii} (i = 1,2), so that

$$E(z_i) = \mu_i \; ; \; E[(z_i - \mu_i)(z_i - \mu_i)'] = \Sigma_{ii} \quad (i = 1,2)$$

Also, we denote the covariance between them as Σ_{12}, in the sense that

$$E[(z_1 - \mu_1)(z_2 - \mu_2)'] = \Sigma_{12}$$

This joint normal distribution can be expressed compactly as

$$\begin{pmatrix} z_1 \\ z_2 \end{pmatrix} \sim N\left[\begin{pmatrix} \mu_1 \\ \mu_2 \end{pmatrix} ; \begin{pmatrix} \Sigma_{11} & \Sigma_{12} \\ \Sigma_{21} & \Sigma_{22} \end{pmatrix}\right] \qquad [2.22]$$

Then, it is well known (see, for example, Anderson, 1958) that the distribution of z_1 given z_2 is also multivariate normal, with mean

$$E(z_z|z_2) = \mu_1 + \Sigma_{12}\Sigma_{22}^{-1}(z_2-\mu_2) \qquad [2.23]$$

and covariance matrix

$$Var(z_1|z_2) = \Sigma_{11}-\Sigma_{12}\Sigma^{-1}{}_{22}\Sigma_{21} \qquad [2.24]$$

Consider now the joint distribution of β_t and y_t given information available at time t – 1; that is, given $y_1, y_2, \ldots, y_{t-1}$. By our assumptions, this distribution is multivariate normal, and, using (2.6) – (2.10), it can be expressed in the format (2.22) as

$$\begin{pmatrix} \beta_t \\ y_t \end{pmatrix}_{t-1} \sim N \left[\begin{pmatrix} \Phi\beta(t-1\,|\,t-1) + (I-\Phi)\beta \\ x_t'\beta(t\,|\,t-1) \end{pmatrix} ; \begin{pmatrix} P(t\,|\,t-1) & P(t\,|\,t-1)x_t \\ x_t'P(t\,|\,t-1) & h_t \end{pmatrix} \right] \qquad [2.25]$$

Now, we require the distribution of β_t given all information available at time t; that is given $y_1, y_2, \ldots, y_t$. This is obtained by conditioning the distribution of β_t at time t – 1 or that of y_t at time t–1, this device efficiently allowing us to introduce the additional conditioning variable y_t. The distribution required is, therefore, multivariate normal, and, identifying terms in (2.25), it follows from (2.23) that its mean is

$$\beta(t\,|\,t) = \Phi\beta(t-1\,|\,t-1) + (I-\Phi)\beta$$

$$+ P(t\,|\,t-1)x_t h_t^{-1}\,[y_t - x_t'\beta(t\,|\,t-1)] \qquad [2.26]$$

and from (2.24) that its covariance matrix is

$$P(t\,|\,t) = P(t\,|\,t-1) - P(t\,|\,t-1)x_t h_t^{-1}\,x_t'P(t\,|\,t-1) \qquad [2.27]$$

Equations (2.26) and (2.27) are (2.11) and (2.12), as required.

3. SOME TESTS OF HYPOTHESES

In this chapter we explore some of the issues involved in testing hypotheses about the stochastic parameter regression model. Our discussion will be based on the simple version of the model, involving just a single independent variable. This allows us to keep the discussion relatively simple, and also to employ, in Section 3.3, a particular test whose extension to the case of more than a single stochastic parameter is very difficult. We will specify a linear regression model, with a fixed intercept term α, and stochastic slope parameter β_t. The model, then, is

$$y_t = \alpha + \beta_t x_t + e_t \quad [3.1]$$

where e_t is an error term having zero mean, fixed variance σ_e^2 and being non-autocorrelated. The single stochastic parameter is again allowed to evolve through time according to a first-order autoregressive process.

$$\beta_t - \beta = \phi(\beta_{t-1} - \beta) + a_t \quad [3.2]$$

Here β is the mean of the stochastic parameter, and the unknown scalar ϕ is the autoregressive coefficient, measuring the correlation between stochastic parameters in adjacent time periods, so that

$$\text{Corr.}(\beta_t, \beta_{t-j}) = \phi^j \quad (j = 1,2, \ldots)$$

This parameter is restricted to take values in the range $-1 < \phi < 1$. In (3.2), the error term a_t has zero mean, fixed variance σ_a^2, and is neither autocorrelated nor correlated with the error terms in (3.1).

We now consider, in increasing order of generality, three specific models in the general class (3.1)–(3.2):

The fixed parameter model. An important special case of our model arises when the regression slope is not stochastic, but rather is fixed through time, so that we can write

$$\beta_t = \beta$$

for all t. This, of course, is the usual linear regression model, for which traditional least squares estimation, and the associated inferential methodology, is appropriate. In terms of our general specification, the error term a_t in (3.2) is then identically zero, or, equivalently, the variance σ_a^2 is zero. The fixed parameter model has, within our more general set-up, a further interesting feature that influences the appropriate hypothesis testing methodology. Once the regression slope is taken to be fixed, the autoregressive coefficient ϕ in (3.2) is irrelevant. As there is no stochastic parameter variation, it does not matter what value is attached to this coefficient, which is therefore said to be *undefined* within the context of the fixed parameter model.

The random coefficient model. Consider now the situation in which the regression slope is stochastic, but not autocorrelated. This is the case in which the autoregressive coefficient ϕ of (3.2) is equal to zero. In that circumstance, the regression slope obeys, through time, the simple stochastic process

$$\beta_t = \beta + a_t$$

It follows that future values of the regression slope are unpredictable, given past information, in the sense that the optimal forecast of any future value is simply the mean β. In this model, then, the regression slope varies randomly through time around its mean, β, according to a distribution whose standard deviation is σ_a. If this behavior is exhibited, the data are said to be generated by a *random coefficient model.*

The first-order autoregressive stochastic parameter model. We will refer to our general model (3.1) and (3.2) in which the autoregressive parameter ϕ is not necessarily zero, as a first-order autoregressive stochastic parameter model. As discussed in Chapter 1, we will restrict attention to the stationary case, in which the autoregressive parameter is constrained to be less than one in absolute value. The random coefficient model is, of course, the special case of the first-order autoregressive stochastic parameter model, in which ϕ is equal to zero.

Given data, an analyst is likely to be interested in trying to determine which of the above three models might be an appropriate generating process. One approach to this problem is to compare the models through hypothesis tests, in which a simpler model is tested against a more elaborate alternative. In this chapter we will discuss and illustrate three such tests.

TESTING THE RANDOM COEFFICIENT MODEL AGAINST THE FIRST-ORDER AUTOREGRESSIVE STOCHASTIC PARAMETER MODEL

Assuming for the time being that the regression slope parameter is stochastic, it is of interest to ask whether future values of this parameter are predictable, or whether values of the stochastic parameter are generated by a purely random process. Specifically, then, we will want to test the null hypothesis

$$H_0: \phi = 0$$

where ϕ is the autoregressive coefficient in equation (3.2), against the alternative hypothesis

$$H_1: \phi \neq 0$$

Hence, our null hypothesis is the random coefficient model, and the alternative is that the stochastic parameter obeys a first-order autoregressive process, with non-zero autoregressive coefficient.

In the following section, we will see how the maximum likelihood estimation procedure of Section 2.2 can be exploited in the development of a likelihood ratio test for this problem.

TESTING THE FIXED PARAMETER MODEL AGAINST THE RANDOM COEFFICIENT MODEL

There are considerable advantages in being able to adequately represent data by a regression model with fixed parameters. Computational costs are very low, and algorithms for the calculation of the usual inferential statistics are universally available. Therefore, before embarking on the tedious work of analyzing a more elaborate formulation, it is natural for the analyst to ask whether such a course is necessary. Accordingly, when contemplating the possibility of stochastic parameter regression models, it is important to have available procedures for testing the adequacy of the fixed parameter model against such alternatives.

The most straightforward possibility of this sort is to test the null hypothesis of a fixed parameter model against the alternative of a random coefficient model. Thus, in terms of our notation, we want to test the null hypothesis

$$H_0: \sigma_a^2 = 0$$

where σ_a^2 is the variance of the error term a_t of (3.2) against the alternative

$$H_1: \sigma_a^2 > 0 \; (\phi = 0)$$

One difficulty associated with this problem is that the null hypothesis specifies a value on the boundary of the parameter surface—that is to say, we know that the variance σ_a^2 cannot be less than zero. In such circumstances, it is well known that the usual inferential procedures based on likelihood ratio tests break down. However, the Lagrange multiplier test remains applicable. (These issues are discussed by Chant, 1974). Accordingly, we will employ the Lagrange multiplier test, introduced in Chapter 1, for this problem.

Now, the fixed parameter model, the random coefficient model, and the more general first-order autoregressive stochastic parameter model constitute a sequence of *nested alternatives,* in the sense that each generalizes the previous model. Accordingly, we could proceed by applying in sequence the two tests just discussed. First, the null hypothess of a random coefficient model can be tested against the alternative in which the autoregressive parameter is not zero. Next, if this null hypothesis is not rejected, the fixed parameter model is tested against the random coefficient model.

Alternatively, at lower computational cost, it may be preferable to test directly the fixed parameter model against the stochastic parameter model whose autoregressive coefficient is not necessarily assumed to be zero.

TESTING THE FIXED PARAMETER MODEL AGAINST THE FIRST-ORDER AUTOREGRESSIVE STOCHASTIC PARAMETER MODEL

In the notation used earlier, our problem is to test the null hypothesis

$$H_o: \sigma_a^2 = 0$$

against the alternative that the regression slope is stochastic, following some unspecified first-order autoregressive model. Hence, our alternative hypothesis is

$$H_1: \sigma_a^2 > 0 \quad (\phi \text{ unspecified in range } -1 < \phi < 1)$$

This particular problem involves, as discussed earlier, the additional difficulty that under the null hypothesis the autoregressive coefficient ϕ is undetermined. This difficulty is circumvented through a modification of the Lagrange multiplier test, the general principle set out by Davies (1977).

3.1 A Test of the Random Coefficients Model Against the First-Order Autoregressive Stochastic Parameter Model

Suppose now that a stochastic parameter regression model is to be fitted to data on a dependent variable and a single independent variable. The null hypothesis of interest is that the regression slope varies purely randomly about its mean, through time. Alternatively, we contemplate the possibility that the stochastic parameter follows a first-order autoregressive model. Our null hypothesis, then, is that the autoregressive parameter ϕ of (3.2) is equal to zero. We test this null hypothesis against the two-sided alternative that ϕ differs from zero, employing a likelihood ratio test.

In Section 2.2, we saw how the Kalman filter algorithm can be used to compute the likelihood function of the stochastic parameter regression model, hence obtaining maximum likelihood estimates of the fixed coefficients of that model. The likelihood ratio test is based on the difference between the maximized log likelihoods under the alternative and null hypotheses. Let L_0 denote the highest possible value of the likelihood function under the null hypothesis, and L_1 the value of the maximized likelihood under the alternative hypothesis. Because the null hypothesis specifies a value for just a single coefficient, it is known that, for large sample sizes, the statistic

$$LR = 2 (\log L_1 - \log L_0) \qquad [3.3]$$

has, under the null hypothesis, a distribution that is well approximated by the chi-square with one degree of freedom. The test of the null hypothesis that the random coefficient model is appropriate can therefore be carried out in the following steps:

(1) Precisely as discussed in Section 2.2, find the maximum of the likelihood function over the permissible range of the fixed coefficients of the general model—that is, $-\infty < \alpha < \infty$, $-\infty < \beta < \infty$, $\sigma_e^2 > 0$, $\sigma_a^2 > 0$, $-1 < \phi < 1$. Denote by L_1 the value of this maximized likelihood.
(2) Repeat the previous step, maximizing the likelihood function over the permissible range of the fixed coefficients, but now with the autoregressive coefficient ϕ fixed to be zero. Denote by L_0 the value of the maximized likelihood subject to this constraint.
(3) Calculate the test statistic (3.3) and compare this with tabulated values of the chi-square distribution with one degree of freedom in the upper tail. If the value of the test statistic exceeds that of the

> tabulated chi-square distribution, reject at the corresponding significance level the null hypothesis of a random coefficient model against the alternative that the stochastic parameter follows a first-order autoregressive model with non-zero coefficient ϕ.

The likelihood ratio test extends, in an obvious way, to the case where the regression equation contains K stochastic parameters. Then the matrix of autoregressive parameters will contain K^2 elements, so that the appropriate comparison of the test statistic is with tabulated values of the chi-square distribution with K^2 degrees of freedom.

In the case of just a single independent variable, as in (3.1) and (3.2), an alternative to the likelihood ratio test is to compare the estimate of the autoregressive parameter with its estimated standard error, using a simple t-test.

3.2 A Test of the Fixed Parameter Model Against the Random Coefficient Model

We now consider the test of the null hypothesis that the regression slope is fixed through time against the alternative that it is purely randomly distributed about its mean. The Lagrange multiplier test is based on the slope of the logarithm of the likelihood function, evaluated under the null hypothesis, where σ_a^2 of (3.2) is equal to zero.

Combining (3.1) and (3.2), with ϕ equal to zero because the random coefficient model is the alternative hypothesis, we have

$$y_t = \alpha + \beta x_t + e_t + x_t a_t$$

This is a regression model with fixed parameters and an error term $(e_t + x_t a_t)$ that has zero mean, is not autocorrelated, and has variance

$$\mathrm{Var}(e_t + x_t a_t) = \sigma_e^2 + x_t^2 \sigma_a^2$$

The likelihood function—which is simply the joint distribution of the observations—is therefore, for a sample of size T and assuming a normal distribution

$$L = (2\Pi)^{-T/2} \prod_{t=1}^{T} (\sigma_e^2 + x_t^2 \sigma_a^2)^{-½}$$

$$\exp.\left[- \sum_{t=1}^{T} (y_t - \alpha - \beta x_t)^2 / 2(\sigma_e^2 + x_t^2 \sigma_a^2) \right]$$

Taking logarithms, we can write the log likelihood as

$$\log L = -\frac{T}{2}\log(2\pi) - \frac{1}{2}\sum_{t=1}^{T}\log(\sigma_e^2 + x_t^2\sigma_a^2)$$

$$-\frac{1}{2}\sum_{t=1}^{T}\frac{(y_t - \alpha - \beta x_t)^2}{(\sigma_e^2 + x_t^2\sigma_a^2)}$$

Differentiating the log likelihood with respect to σ_a^2 yields

$$\frac{d\log L}{d\sigma_a^2} = -\frac{1}{2}\sum_{t=1}^{T}\frac{x_t^2}{\sigma_e^2 + x_t^2\sigma_a^2} + \frac{1}{2}\sum_{t=1}^{T}\frac{x^2(y_t - \alpha - \beta x_t)^2}{(\sigma_e^2 + x_t^2\sigma_a^2)^2} \qquad [3.4]$$

We now need to evaluate (3.4) under the null hypothesis of a fixed parameter model. Let us assume, then, that the model

$$y = \alpha + \beta x_t + e_t \qquad [3.5]$$

holds. Then, we can estimate (3.5) by ordinary least squares. Let a and b denote the least squares estimates of α and β, denote the residuals from the fitted least square regression by

$$r_t = y_t - a - bx_t \quad (t = 1, 2, \ldots, T)$$

The maximum likelihood estimate of the variance σ_e^2 is, then,

$$s_e^2 = \sum_{t=1}^{T} r_t^2 / T$$

Substituting these estimates, together with the value $\sigma_a^2 = 0$ implied by the null hypothesis, equation (3.4) then yields for the estimated slope of the log likelihood function

$$\frac{d\log L}{d\sigma_a^2} = -\frac{1}{2}\sum_{t=1}^{T}\frac{x_t^2}{s_e^2} + \frac{1}{2}\sum_{t=1}^{T}\frac{x_t^2 r_t^2}{s_e^4}$$

The Lagrange multiplier statistic is then obtained by dividing this quantity by an estimate of its standard deviation under the null hypothesis. In fact, as shown by Breusch and Pagan (1979), who devised this particular test, this yields the test statistic

$$S = \frac{\frac{1}{2}\sum_{t=1}^{T} x_t^2 \left(\frac{r_t^2}{s_e^2} - 1\right)}{\left[\frac{1}{2}\sum_{t=1}^{T} x_t^4 - \frac{1}{2T}\left(\sum_{t=1}^{T} x_t^2\right)^2\right]^{1/2}} \qquad [3.6]$$

The statistic S is employed as the basis of the Lagrange multiplier test of the null hypothesis of a fixed parameter model, when the alternative is a random coefficient model. Under the null hypothesis, this statistic has, for large samples, a distribution that is well approximated by the standard normal. Because the alternative hypothesis is that the variance σ_a^2 is positive, that hypothesis will be preferred to the null for large positive values of the slope of the log likelihood function—that is, for large positive values of the statistic (3.6). The test is therefore carried out as follows:

(1) Estimate by ordinary least squares the simple linear regression (3.5) of y_t on x_t, finding the residuals r_t and the estimated error variance s_e^2.
(2) Compute the test statistic S of (3.6). The null hypothesis of a fixed parameter model is then rejected against the random coefficient alternative at significance level α if S is greater than z_α, where z_α is the number exceeded with probability α by a standard normal random variable.

Let us try to get some further insight into the rationale behind the use of the test statistic (3.6). If the fixed parameter regression model is appropriate, the error terms e_t in the linear regression (3.5) will have a constant variance through time. On the other hand, under the random coefficient model the variance of these errors will be an increasing function of x_t^2. Therefore, if the random coefficient model holds, we would expect to find a positive correlation between x_t^2 and e_t^2. Hence, it would be natural to base a test of the fixed parameter model against the random coefficient alternative on an estimate of this correlation. We now show that this is precisely what is accomplished by the test that we have just described.

Write the correlation between x_t^2 and e_t^2

$$\text{Corr.}\,(x_t^2, e_t^2) = \frac{E(x_t^2 e_t^2) - E(x_t^2)E(e_t^2)}{\sqrt{\text{Var}(x_t^2)\,\text{Var}(e_t^2)}} \qquad [3.7]$$

Now, we estimate the expectation of x_t^2 by $\Sigma x_t^2/T$, and the variance of x_t^2 by

$$\text{Est. Var}(x_t^2) = \frac{1}{T}\left[\Sigma x_t^4 - \frac{1}{T}(\Sigma x_t^2)^2\right]$$

Because the residuals r_t from the fitted regression can be used to estimate the unknown errors e_t, a natural estimate of the expectation of the product $x_t^2 e_t^2$ is provided by $\Sigma x_t^2 r_t^2/T$. Finally, under our null hypothesis, the error terms e_t are normally distributed with mean zero and variance σ_e^2, from which it follows that e_t^2 has expectation σ_e^2 and variance $2\sigma_e^4$. Natural estimates of this mean and variance are given by s_e^2 and $2s_e^4$.

Substituting these estimates into (3.7), we then have as an estimate of the correlation between x_t^2 and e_t^2

$$\text{Est. Corr.}\,(x_t^2, e_t^2) = \frac{\frac{1}{T}\sum_{t=1}^{T} x_t^2 r_t^2 - \frac{1}{T}\sum_{t=1}^{T} x_t^2 s_e^2}{\left[\frac{1}{T}\sum_{t=1}^{T} x_t^2 - \frac{1}{T^2}\left(\sum_{t=1}^{T} x_t^2\right)^2\right]^{1/2}(2s_e^4)^{1/2}} \qquad [3.8]$$

Comparing expressions (3.6) and (3.8), we find that the Lagrange multiplier test statistic S is simply a multiple $T^{1/2}$ of the estimated correlation between X_t and e_t. This provides a strong rationale for the use of the Lagrange multiplier statistic as the basis of a test of the null hypothesis of a fixed parameter model against the random coefficient alternative.

One of the great advantages of this particular test is the relative ease with which it can be carried out. As we noted in our discussion of the Lagrange multiplier test in Chapter 1, the model need only be estimated under the null hypothesis. This is particularly advantageous in terms of computation time in the present context. We have only to estimate the simple linear regression equation (3.5) by least squares to generate all of the information needed to calculate the statistic (3.6). It is not necessary to estimate also the random coefficient model.

Breusch and Pagan (1979) discuss the general case of this test when the regression equation contains several independent variables.

3.3 A Test of the Fixed Parameter Model Against the First-Order Autoregressive Stochastic Parameter Model

The final test to be discussed here has obvious practical desirability. Before embarking on the estimation of a stochastic parameter regression model, one would wish to have evidence that the simple fixed coefficient linear regression model is inadequate. Given that, a priori, there will generally be no strong reasons for believing that any stochastic parameter behavior is of the random coefficient sort, we would want to contemplate as an alternative hypothesis the possibility that the regression slope obeys a first-order autoregressive model, with coefficient ϕ allowed to take any value in the range $-1 < \phi < 1$. As already noted, testing against such an alternative leads to the position where ϕ is undefined under the null hypothesis. The general approach of Davies (1977) to such problems has been applied in the present context by Watson (1980).

First, let us consider the position where the alternative hypothesis specifies *a particular value* ϕ for the autoregressive coefficient. (The case discussed in the previous section is, of course, one in which a zero value is specified for this coefficient). It can then be shown that the Lagrange multiplier statistic for testing the null hypothesis of a fixed parameter model against this alternative is

$$S(\phi) = \frac{\frac{1}{2}\sum_{t=1}^{T} x_t^2 \left[\frac{r_t^2}{s_e^2} - 1\right] + \frac{1}{s_e^2}\sum_{t=2}^{T} r_t x_t \sum_{i=1}^{t-1} r_i x_i \phi^{t-i}}{\left[\frac{1}{2}\sum_{t=1}^{T} x_t^4 + \sum_{t=2}^{T} x_t^4 \sum_{i=1}^{t-1} x_i^2 \phi^{2(t-i)} - \frac{1}{2T}\left(\sum_{t=1}^{T} x_t^2\right)^2\right]^{1/2}} \qquad [3.9]$$

Notice that the statistic S of (3.6) is simply the special case of $S(\phi)$ where the autoregressive parameter is equal to zero. Under the null hypothesis the statistic (3.9) has, to a good approximation in large samples, a standard normal distribution.

Now, our alternative hypothesis is not a stochastic parameter model with some particular value of ϕ, but is rather of such a model where ϕ might take *any* value in the range $-1 < \phi < 1$. This suggests that we might compute the statistic (3.9) for a grid of possible values of ϕ in this region. We would then be suspicious of the null hypothesis if *any* of the computed statistics was unduly large. Therefore, we might base a test of the null hypothesis that the fixed parameter model is adequate on the largest value taken by $S(\phi)$ over the range $-1 < \phi < 1$, that is, on

$$\underset{-1<\phi<1}{\text{Sup}}\ S(\phi) \qquad [3.10]$$

Unfortunately, although the distribution of $S(\phi)$ under the null hypothesis is well approximated in large samples by a standard normal, that of the statistic (3.10) is not. This should not be surprising because that statistic is the supremum of observations drawn from different standard normal distributions. It is, however, possible to find an approximate upper bound to the probability that (3.10) exceeds any given number. Let c be some fixed number. Then it is possible to show that, for large samples,

$$P(\underset{-1<\phi<1}{\text{Sup.}}\ S(\phi) \geqslant c) \leqslant F(-c) + U(c) \qquad [3.11]$$

where F is in the cumulative distribution function of the standard normal distribution, so that

$$F(-c) = \int_{-\infty}^{-c} (2\Pi)^{-1/2} \exp(-z^2/2)\, dz$$

and U(c) is defined by

$$U(c) = (2\Pi)^{-1} \exp(-c^2/2) \int_{-1}^{1} (-\rho_{11}(\theta))^{1/2} d\theta \qquad [3.12]$$

where

$$\rho_{11}(\theta) = \left[\frac{\partial^2}{\partial \phi^2}\, \rho(\phi, \theta) \right]_{\phi=\theta} \qquad [3.13]$$

That is, $\rho_{11}(\theta)$ is the second partial derivative of $\rho(\phi,\theta)$ with respect to ϕ, evaluated at $\phi = \theta$. Here, $\rho(\phi,\theta)$ is defined as

$$\rho(\phi,\theta) = \frac{\frac{1}{2}\sum_{t=1}^{T} x_t^4 + \sum_{t=1}^{T} x_t^2 \sum_{i=1}^{t-1} x_i^2 \phi^{t-i}\theta^{t-i} - \frac{1}{2T}\left(\sum_{t=1}^{T} x_t^2\right)^2}{\lambda(\phi)\lambda(\theta)} \qquad [3.14]$$

where

$$\lambda(\gamma) = \left[\frac{1}{2}\sum_{t=1}^{T} x_t^4 + \sum_{t=2}^{T} x_t^2 \sum_{i=1}^{t-1} x_i^2 \gamma^{2(t-i)} - \frac{1}{2T}\left(\sum_{t=1}^{T} x_t^2\right)^2\right]^{1/2} \qquad [3.15]$$

In practice, the partial derivative (3.13) is best found numerically, rather than through analytical differentiation of (3.14). The integral of (3.12) can also be easily found by numerical methods. The test procedure is then as follows:

(1) Estimate by ordinary least squares the simple linear regression of y_t on x_t, obtaining the residuals r_t and the estimate s_e^2 of the error variance.
(2) Calculate the statistic (3.9) for a grid of possible values of ϕ, and hence locate the highest value $S(\phi)$ takes in the range $-1 < \phi < 1$. Denote this supremum by c.
(3) The quantity F(–c) in (3.11) can be read directly from tables of the cumulative distribution function of the standard normal distribution, whereas U(c) is found from equations (3.12) – (3.15).
(4) The null hypothesis that the fixed parameter regression model is adequate can then be rejected against the alternative of a first order autoregressive stochastic parameter model, with unspecified ϕ, at any significance level greater than F(–c) + U(c).

The quantity U(c) of (3.11) can be viewed as a correction to the significance level implied by the standard normal distribution, this correction being necessitated by our search for the supremum of the statistics (3.9). The need for this correction factor certainly increases somewhat the computational burden involved in carrying out the test. Nevertheless, in total the amount of computing time required remains relatively modest, as only the fixed parameter regression model has to be estimated. It is not necessary to estimate the stochastic parameter model in order to carry out this test.

3.4 Application to the Market Model

In this section we discuss the application of the methodology of the present and previous chapters to a model that has been widely applied in the finance literature. The *capital asset pricing model,* developed by Sharpe (1964) and Lintner (1965a, 1965b), and discussed by Fama (1968), deals with the equilibrium rate of return on risky securities. Rate of return over a period of time is the sum of the price change and dividends received over the period, expressed as a proportion of the price at the beginning of the period. Under certain idealized assumptions, if R_i denotes the rate of return of the ith security, then, according to the capital asset pricing model, the expected rate of return of this security can be expressed as

$$E(R_i) = R_F + \beta_i[E(R_M) - R_F]$$

where R_F is the risk-free rate of interest and $E(R_M)$ is the expected rate of return on a market portfolio of securities, consisting of every asset outstanding in proportion to its total value. The quantity β_i is given by

$$\beta_i = Cov(R_i,R_M)/Var(R_M)$$

and is known as the *systematic risk* of the ith security.

One procedure for estimating systematic risk is to obtain time series R_{it} and R_{Mt} of observations on rates of return of the security of interest and of the market portfolio, and estimate by ordinary least squares the linear regression model

$$R_{it} = \alpha_i + \beta_i R_{Mt} + e_{it} \qquad [3.16]$$

where α_i is an intercept term, and e_{it} is an error term, generally taken to have mean zero, fixed variance through time, and to be nonautocorrelated. The formulation (3.6) is generally known in the finance literature as the *market model.* One interpretation of this model follows from writing

$$Var(R_{it}) = \beta_i^2 Var(R_{Mt}) + Var(e_{it})$$

The variability, or risk, associated with the rate of return on investment in the ith security is decomposed into two parts. The first of the components shows how variability in the market rate of return relates to variability in the rate of the individual security. We see that the higher the β_i, the greater the impact of market variability on variability of returns on the security. It is for this reason that β_i is referred to as "systematic risk."

The systematic risk of a security, then, provides an important ingredient in an investor's decision-making process. Estimation of this parameter by the application of the ordinary least squares to equation (3.16) effectively implies a belief that systematic risk remains fixed through time. However, a number of authors, including Rosenberg and Guy (1976a, 1976b) have argued that it would be more plausible to expect systematic risk to vary through time. This variation might arise as a result of microeconomic factors, such as changes in the operational structure of a corporation or in its business environment, or macroeconomic factors, such as general business conditions, expectations about relevant future events, or inflation. Some support for the contention that systematic risk may vary through time is given in Jacob (1971), Blume (1975), and Fabozzi and Francis (1978).

We consider, then, the possibility that in the market model systematic risk is stochastic, generated perhaps by a first-order autoregressive model. Therefore, corresponding to (3.1) and (3.2), we consider the first-order autoregressive stochastic parameter model.

$$\begin{aligned} R_{it} &= \alpha_i + \beta_{it} R_{Mt} + e_{it} \\ \beta_{it} - \beta_i &= \phi_i(\beta_{i,t-1} - \beta_i) + a_{it} \end{aligned} \qquad [3.17]$$

The random coefficient model—which is of course the special case of (3.17) with the autoregressive coefficient ϕ_i equal to zero—has been employed in analyzing the market model by Fabozzi and Francis (1978), Lee and Chen (1980), Alexander and Bensen (1982), and Fabozzi et al., (1982). The more general first-order autoregressive stochastic parameter model has been considered by Sunder (1980), Ohlson and Rosenberg (1982), and Bos and Newbold (1984).

We will illustrate our methodology by considering a sample of monthly observations over a period of ten years, running from January 1970 to December 1979. To data on ABT Abbot Labs, we fitted by ordinary least squares the fixed parameter regression model, and by the maximum likelihood method of Chapter 2, the random coefficient model and the first-order autoregressive stochastic parameter model, obtaining the following results:

(1) The fixed parameter model. The estimated model obtained was

$$R_t = \underset{(0.00506)}{0.00266} + \underset{(0.105)}{1.057}\, R_{Mt} + e_t \ ; \qquad s_e^2 = 0.00303$$

where here and throughout, figures in parentheses beneath coefficient estimates are the corresponding estimated standard errors, and s_e^2 is the estimate of the error variance σ_e^2. The coefficient of determination, R^2, was 0.456. The Durbin-Watson statistic, 1.700, suggests no great concern about autocorrelated errors, and the maximized log likelihood is 177.749.

(2) The random coefficient model. Estimating by maximum likelihood the model (3.17), with the autoregressive coefficient forced to take the value zero, we found

$$R_t = \underset{(0.00480)}{0.00368} + \beta_t R_{Mt} + e_t \ ; \qquad s_e^2 = \underset{(0.00040)}{0.00227}$$

$$\beta_t - \underset{(0.122)}{0.958} = a_t \ ; \qquad s_a^2 = \underset{(0.182)}{0.294}$$

where s_a^2 is the estimate of the error variance a_t. The maximized log likelihood was found to be 182.070.

(3) The first-order autoregressive stochastic parameter model. When the full model (3.17) was estimated by maximum likelihood, we obtained

$$R_t = \underset{(0.00480)}{0.00409} + \beta_t R_{Mt} + e_t \ ; \qquad s_e^2 = \underset{(0.00040)}{0.00227}$$

$$\beta_t - \underset{(0.129)}{0.945} = \underset{(0.441)}{0.274}(\beta_{t-1} - \underset{(0.129)}{0.945}) + a_t \ ; \qquad s_a^2 = \underset{(0.189)}{0.265}$$

The maximized log likelihood was 182.389.

We can interpret the point estimates of these three models. The implication of the fixed parameter model is that systematic risk is estimated to be 1.057 throughout this ten-year period. The random coefficient model implies that systematic risk varies randomly through time around a mean of 0.958. For the first-order autoregressive stochastic parameter model, we estimate that systematic risk varies around a mean of 0.945 according to a first-order autoregressive model with coefficient 0.274. Hence we estimate a correlation of 0.274 between systematic risks of this security in adjacent months. This finding can be interpreted as implying that, standing at time T, the optimal forecast of the future systematic risk β_{T+j} is

$$0.945 + (0.274)^j(\beta_T - 0.945)$$

We now apply to these data the three hypothesis tests discussed earlier in this chapter.

TESTING THE RANDOM COEFFICIENT MODEL AGAINST THE FIRST-ORDER AUTOREGRESSIVE STOCHASTIC PARAMETER MODEL

In terms of the model (3.17) we require to test the null hypothesis

$$H_0: \phi = 0$$

against the alternative

$$H_1: \phi \neq 0$$

Recalling our discussion of Section 3.1, the test is based on the likelihood ratio statistic (3.3) We found that, under the alternative hypothesis, the maximized log likelihood is

$$\log L_1 = 182.389$$

However, for the random coefficient model, the maximized log likelihood is

$$\log L_0 = 182.070$$

The likelihood ratio statistic (3.3) is therefore

$$LR = 2(\log L_1 - \log L_0) = 2(182.389 - 182.070) = 0.638$$

Comparing this statistic with tabulated value of the chi-square distribution with one degree of freedom, we find that the null hypothesis cannot be rejected in favor of the alternative at the usual significance levels. Hence these data do not provide strong evidence to suggest that, if a stochastic parameter model is appropriate, the autoregressive coefficient of that model is other than zero.

Notice that we could have reached a similar conclusion by a rather different route. The point estimate of the autoregressive coefficient of the first-order autoregressive stochastic parameter model is 0.274. However, the associated standard error, 0.441, is extremely large. Hence, because the estimator has a distribution that is close to normal in large samples, we conclude that the estimated autoregressive coefficient does not differ significantly from zero.

TESTING THE FIXED PARAMETER MODEL AGAINST THE RANDOM COEFFICIENT MODEL

Next we want to test the null hypothesis

$$H_0: \sigma_a^2 = 0$$

against the random coefficient alternative

$$H_1: \sigma_a^2 > 0 \quad (\phi = 0)$$

using the Lagrange multiplier test of Section 3.2. For these particular data, the test statistic S of equation (3.6) was found to be 5.438. If the null hypothesis was correct, this figure would represent a random drawing from a standard normal distribution. It follows then that the null hypothesis is very clearly rejected at the usual significance levels. The evidence against the fixed parameter model is extremely strong.

This approach is, in our experience, superior to the comparison of the estimate of σ_a^2 with its standard error, a procedure that in this context is invalid and unreliable.

TESTING THE FIXED PARAMETER MODEL AGAINST THE FIRST-ORDER AUTOREGRESSIVE STOCHASTIC PARAMETER MODEL

Finally, using the Davies-Watson test, we test the null hypothesis

$$H_0: \sigma_a^2 = 0$$

against the alternative hypothesis

$$H_1: \sigma_a^2 > 0 \quad (\phi \text{ unspecified in range } -1 < \phi < 1)$$

using the procedure outlined in Section 3.3. The statistic $S(\phi)$ of (3.9) was computed for values of ϕ ranging from –0.99 to 0.99, in increments of 0.01. The largest value of this statistic, occurring at $\phi = 0.39$, was 6.0560. Setting $c = 6.0560$, we found the corresponding value of (3.11) to be

$$F(-6.0560) + U(6.0560) = 0.940 \times 10^{-8}$$

It follows, then, that the null hypothesis that the fixed parameter regression model is adequate can be rejected at any significance level greater than 0.940×10^{-8}. The evidence against the null hypothesis is overwhelming.

Bos and Newbold (1984) analyzed in this way data on a sample 464 stocks traded on the New York Stock Exchange. The results on ABT Abbot Labs are rather typical of our findings. The fixed parameter model was rejected against both the random coefficient model and the first-order autoregressive stochastic parameter model on a clear majority of occasions using 5% significance level tests. The available evidence, then, argues strongly against the general adequacy of the simple fixed parameter market model (3.16). On the other hand, only for a very few securities was the null hypothesis of a random coefficient model rejected against the alternative that the autoregressive coefficient differs from zero. We failed to find strong evidence against the proposition that the random coefficient model is appropriate. This should not, however, be taken as strong evidence *in support of* the random coefficient market model. Indeed, the point estimates of the autoregressive coefficients were widely dispersed over the range $-1 < \phi < 1$. However, just as in the case of the ABT Abbot Labs data, the associated standard errors were generally very large. This leads us to suspect that, for these data, the test of the random coefficient model is not very powerful. Nevertheless, as a working hypothesis for further investigation, our study suggests the attraction of the random coefficient model.

4. TESTING FOR EFFICIENT CAPITAL MARKETS

There are many conceivable applications of stochastic parameter models in econometrics. Most applications can be regarded as attempts to model dynamic structural change in a system. In this chapter an economic hypothesis is tested. In particular, stochastic parameter regression models are used to test for the economic efficiency of the market for 3-month U.S. Treasury bills (T-bills).

Fama (1975) was interested in finding out whether the market for short term T-bills is efficient. His work provided the incentive for a number of dissenting studies that Fama addressed in his 1977 article. In his review of the literature of efficient capital markets, Fama (1970) defines an efficient market to be one in which prices fully reflect all available information. In the case of capital markets this means that an efficient capital market is one in which the nominal rate of interest on a bond is set, by the market, to equal the equilibrium real return plus the fully anticipated rate of inflation. The only type of bond for which investors can accurately predict its price at the end of a period, and hence perfectly forecast its rate of return, is one that matures at the end of the period. Hence, if the market for short term T-bills is efficient, then

$$I_t = -r_t + R_t \qquad [4.1]$$

where

R_t is the nominal interest rate on a maturing one period T-bill at the beginning of period t;

r_t is the equilibrium real interest rate at the end of period t;

and

I_t is the rate of inflation during period t.

In contrast to this, in a world of uncertainty, the best the market could do is to set the nominal interest rate on a maturing bond equal to the market's expected equilibrium real rate of return plus the market's expectation of the rate of inflation over the period.

4.1 Fama's Method of Testing for an Efficient T-Bill Market

To test whether a short term T-bill market is efficient, Fama (1975) applied the regression

$$I_t = \alpha + \beta R_t + e_t \qquad [4.2]$$

to a set of contemporary data. If the particular short-term capital market is efficient then equations (4.1) and (4.2) are the same. Thus, if $\beta = 1$ and $e_t + \alpha = -r_t$, then the capital market is efficient. Whether or not β is significantly different from unity can be checked statistically. The other condition relies on the assumption that the equilibrium real rate of return is constant through time, or more specifically, constant plus a white noise error. Fama tests these two conditions in two tests; first a test on β, whose estimate is found to be insignificantly different from unity; and then second, r_t, the equilibrium real rate of return is calculated for estimated using the formula

$$r_t = R_t - I_t \qquad [4.3]$$

The constancy of this series of r_t is tested and confirmed using time-series methods. Thus Fama concludes that the short-term T-bill market seems to be efficient.

What he is actually doing is to first assume that the equilibrium real rate of return is constant through time, except perhaps for a white noise error term, and to test whether the short term T-bill market is efficient. Then once the T-bill market is "shown" to be efficient after assuming the equilibrium real rate of return to be constant, the equilibrium real rate of return is calculated under the assumption of an efficient market and subsequently found to be constant through time. This argument is suspect because it depends on the validity of the first assumption. Because both hypotheses are related, it is preferable that they be tested simultaneously.

Instead of running the linear regression as in equation (4.2), if one estimates the model

$$I_t = \alpha_t + \beta_t R_t + e_t \qquad [4.4]$$

where α_t and β_t are allowed to vary through time, then theoretically one could test both hypotheses at once. Actually the only issue of importance was, and still is, whether or not the short-term T-bill markets are efficient. So it does not matter at all to the question of market efficiency if equilibrium real rate of return changes through time or remains constant. Nevertheless, both alternatives must be allowed, and this is exactly what equation (4.4) does. The next step involves the choice of how α_t and β_t are allowed to vary through time, and this must necessarily be arbitrary. Before making a decision, it is better that these two quantities be described. α_t is the negative of the market's expectation of equilibirum real rate of return, regardless of whether the market is correct or not. β_t can be thought of as an adjustment factor that is either equal to one when the market is efficient, or converging to one, as the inefficient market's participants obtain more information or as they try to correct their intuition for some exogenous white noise. It is somewhat inconceivable that α_t or β_t will change dramatically from one time period to the next. That is to say that α_t and β_t will, for the most part, change rather smoothly through time. Thus it should be possible for the relationship between inflation and the nominal interest rate on a maturing one-period T-bill to be represented by a stochastic parameter model.

4.2 The Data

Because Fama's actual data are not readily available, the nonseasonally adjusted Consumer Price Index (CPI) data were taken from the U.S. Department of Labor, Bureau of Labor Statistics, and the rate on T-bills was obtained for 3-month T-bills from the Board of Governors of the Federal Reserve System. It was decided to test the efficiency of the 3-month T-bill market rather than of the one-month T-bill market that Fama examined, because a monthly data set over two or more decades would involve a very heavy computational burden in finding the maximum likelihood estimates (MLEs) of a stochastic parameter regression model (SPRM).

Although monthly data are available for both the CPI and the 3-month T-bill rate, the two data series were transformed to quarterly data to shorten the data set and to make the unit time period three months to coincide with the length of the T-bill. The quarterly figure for the rate on 3-month T-bills is the average of the three one-month rates. As the CPI is calculated as the average for one month, a quarterly index was calculated by taking the geometric mean of the three monthly figures. Then, to derive the series of inflation rates at the end of each quarter, the percentage change between the end and beginning of each quarter was calculated. More explicitly,

> R_t is the average rate on 3-month T-bills in quarter t, so that it is an estimate of the rate in the middle of quarter t;

and

> I_t is the average rate of inflation from mid-quarter t to mid-quarter t + 1.

In his 1975 study of efficiency of the one-month T-bill market, Fama used monthly data from January 1953 to July 1971. He chose this particular timeframe because interest rates on T-bills were pegged from during World War II until the end of 1952, the calculation of the Consumer Price Index changed substantially at the beginning of 1953, and the Consumer Price Index figures between August 1971 and mid-1974 are suspect, as price controls were in effect.

Here we choose to begin our data set in the first quarter of 1953 (1953/1). To follow the spirit of Fama's study, we at first ended the data set in the second quarter of 1971 (1971/2). Because of what was found as we analyzed the data, we will show statistical results on a series of six data sets each beginning in 1953/1 but ending at three-year intervals starting at the second quarter of 1965 (1965/2)—that is, from 1953/1 to 1965/2, 1968/2, 1971/2 (corresponding to Fama's time frame), 1974/2, 1977/2, and 1980/2.

4.3 Estimation of Stochastic Parameter Regression Models

The procedure for testing for a SPRM, as described in Section 3.3, is meant only for the case in which there is only one parameter that follows an AR(1) process. The generalization to tests against a vector AR(1) SPRM can be derived, as explained by Watson (1980), but this is computationally extremely expensive. Hence a heuristic approach is used to gain some insight as to whether or not the 3-month T-bill market is efficient. This approach progresses in a stepwise manner to estimate and test increasingly more involved stochastic parameter regression models.

Estimation of the ordinary linear regression model. First, all the maximum likelihood estimates are found for the ordinary linear regression model,

$$I_t = \alpha + \beta R_t + e_t \qquad [4.5]$$

where e_t is an error term having zero mean, fixed variance σ_e^2, and is non-autocorrelated. Let a, b, and s_e^2 denote the maximum likelihood estimates of α, β and σ_e^2. These results are displayed in Table 4.1. Although these results are to be used a little later for purposes of comparison, they can also be compared to Fama's least squares estimates for his data on the one-month T-bill market. It is interesting to note that b is statistically insignificantly different from one for the time frames that are equal to or shorter in length than Fama's time frame; b is significantly different from one in the three longer time frames. We note that Fama suggested that the T-bill markets could be inefficient from the middle of 1971 to the middle of 1974 as the data was potentially

TABLE 4.1
Maximum Likelihood Estimates of the Ordinary Linear Regression Model (Equation 4.5)

1953/1 to	*a (SE)*	*b (SE)*	s_e^2	*DW*	*Log Likelihood*
1965/2	−.3489 (.5760)	.6775 (.2118)	1.6891	1.4608	−84.052
1968/2	−.6210 (.4562)	.8089 (.1434)	1.6619	1.4282	−103.722
1971/2	−.9351 (.3539)	.9439 (.0934)	1.5910	1.4618	−122.184
1974/2	−1.8625 (.3888)	1.2781 (.0931)	2.3734	1.0578	−159.194
1977/2	−1.9911 (.3965)	1.3528 (.0901)	2.6381	1.1037	−186.589
1980/2	−2.0799 (.3455)	1.3907 (.0680)	2.8377	1.0843	−213.449

contaminated by governmental price controls. Still, even if the contaminated data are deleted from the estimation of equation (4.5), one would find that b is still significantly different from one for the longest time frame. After looking at the figures for inflation, it seems that what may cause β to be different from one are periods of high, runaway inflation. If this is so, then T-bill market efficiency only seems to occur in stable or more predictable eras. Although this seems reasonable, it does not bode well for a strong test of efficiency in short term T-bill markets.

Estimation of a random coefficient model. Secondly, the maximum likelihood estimations of the simplest kind of stochastic parameter model are calculated. This is the random coefficient model (RCM)

$$I_t = \alpha + \beta_t R_t + e_t \qquad [4.6a]$$

$$\beta_t = \beta + a_t \qquad [4.6b]$$

where e_t is an error term having zero mean, fixed variance σ_e^2, and is non-autocorrelated, and where a_t has a zero mean, fixed variance σ_a^2, and is neither autocorrelated nor correlated with the error term in (4.6a). In this stochastic parameter model, both the intercept and slope are white

TABLE 4.2
Maximum Likelihood Estimates of the Random Coefficients Model (Equation 4.6)

1953/1 to	a (SE)	b (SE)	s_e^2 (SE)	s_a^2 (SE)	S	Log Likelihood
1965/2	−.5946 (.5273)	.7790 (.2057)	1.1350 (.5651)	.077036 (.080218)	.5378	−83.811
1968/2	−.6451 (.4471)	.8179 (.1451)	1.4395 (.4984)	.022062 (.045019)	.4679	−103.608
1971/2					−.6714	
1974/2	−1.4277 (.3409)	1.1526 (.0989)	1.0258 (.3842)	.076949 (.031867)	3.4774	−154.388
1977/2	−1.6215 (.3385)	1.2518 (.0933)	1.0882 (.4011)	.083699 (.030531)	2.7190	−182.374
1980/2	−1.8585 (.3170)	1.3337 (.0777)	1.3299 (.4066)	.064371 (.023560)	2.2213	−209.590

noise. In line with the notation above, let a, b, s_e^2, and s_a^2 be the maximum likelihood estimates of the parameters of the RCM parameters, α, β, σ_e^2, and σ_a^2. The results of this estimation are displayed in Table 4.2.

First, note that the estimates for the RCM could not be found for the time frame ending in 1971/2, the time frame that Fama used. The S test statistics for a RCM were calculated with the maximum likelihood estimations of the fixed parameter regression model, equation (4.5), shown in Table 4.1. For a description of the test for a fixed parameter regression model against the alternative of a RCM, see Section 3.2. This inability to estimate a RCM will occur every time S is negative. This is because S is an estimate of the first derivative of the log-likelihood with respect to σ_a^2, when $\sigma_a^2 = 0$. This is a signal that the log-likelihood function of the RCM decreases from its open boundary with the general linear model (4.5), which is the same as saying that the maximum likelihood estimation of σ_a^2 for the RCM is negative, and this is impossible.

With a significance level of 5%, the critical value of the test statistic S is +1.645 as the test is one-tailed. Thus the null hypothesis of a fixed parameter model cannot be rejected if the S statistic, calculated from that fixed parameter regression model, is less than or equal to 1.645. Also, if the S statistic is greater than 1.645, then one will reject the null hypothesis and accept the alternate of a RCM. Note that the RCM is indicated at this significance level only for the last three time frames—again information that points to the possibility that an inefficient 3-

month T-bill market occurs in the data not used by Fama. The high level of significance of the RCM in the last three time frames may suggest that a more general stochastic parameter model may be more applicable to the entire data set.

Estimation of a stochastic parameter regression model with the slope following a first-order autoregressive process. The next step is to allow β_t to follow an AR(1) process,

$$I_t = \alpha + \beta_t R_t + e_t \qquad [4.7a]$$

$$(\beta_t - \beta) = \phi(\beta_{t-1} - \beta) + a_t \qquad [4.7b]$$

where e_t is an error term having zero mean, fixed variance σ_e^2 and is non-autocorrelated, and where a_t has zero mean, fixed variance σ_a^2, and is neither autocorrelated nor correlated with the error term in (4.7a). Let a, b, s_e^2, s_a^2, and p be the maximum likelihood estimates of the parameters of the SPRM (4.7), α, β, σ_e^2, σ_a^2, and ϕ. The MLEs of model (4.7) and associated test statistics are displayed in Table 4.3. Watson's test for a SPRM with only one slope coefficient following an AR(1) process is shown in the columns headed by p_s and c. p_s is the value of ϕ that maximizes the value of the test statistic $S(\phi)$ as given in equation (3.9). Details of this test—a test for a fixed parameter regression model against the model in which the one slope coefficient follows an AR(1) process—are shown in Section 3.3. The figures beneath c are probabilities that show how likely it is that the slope coefficient is constant through time vis-à-vis an alternative hypothesis that the slope coefficient follows an AR(1) process.

When testing the null hypothesis of a fixed parameter regression model against the alternative of a single slope coefficient following an AR(1) process, we find that, again, the hypothesis of a fixed slope is soundly rejected in the last three time frames. Now, surprisingly there is a high significance for the stochastic slope in the first time frame. So, the only time frames in which a significant SPRM was not found are the second and third—the third corresponding to Fama's time frame. These estimates of model (4.7) for the second and third time frames, viewed by themselves, would lead one to conclude that model (4.7) is not necessary—but when viewed with the results of the other time frames, the insignificance in the second and third time frames appears to be a fluke, especially when one considers the remarkable similarity of the estimated parameters in each time frame. It is, of course, impossible that β_t follows

TABLE 4.3
Maximum Likelihood Estimates of the Stochastic Parameter Regression Model (Equation 4.7)

1953/1 to	a (SE)	b (SE)	s_e^2 (SE)	s_a^2 (SE)	p (SE)	p_s	c (prob)	Log Likelihood (LR test)
1965/2	−.7715 (.6033)	.8584 (.2591)	1.2782 (.3436)	.016097 (.026992)	.84143 (.21220)	.75	2.5812 (.0354)	−81.95 (3.725)
1968/2	−.7102 (.5210)	.8583 (.1964)	1.2237 (.3224)	.018183 (.024438)	.78164 (.23136)	.68	1.9324 (.1518)	−101.616 (3.982)
1971/2	−.7059 (.4636)	.8865 (.1624)	1.2516 (.2594)	.007112 (.009050)	.87224 (.13596)	.79	1.4775 (.3347)	−119.994
1974/2	−.9869 (.5037)	1.1018 (.2073)	1.2102 (.2378)	.010320 (.007840)	.92983 (.06645)	.75	9.7005 (.0000)	−143.352 (22.072)
1977/2	−.9007 (.5208)	1.0876 (.1980)	1.3568 (.2566)	.012248 (.008272)	.91650 (.06503)	.83	9.7987 (.0000)	−169.919 (24.912)
1980/2	−.9705 (.4869)	1.0903 (.1648)	1.4176 (.2666)	.013738 (.008100)	.89636 (.06377)	.85	7.0374 (.0000)	−195.918 (27.342)

an AR(1) process in the first 50, and then the first 86, 98, and 110 observations, but not on the first 62 or 74 observations. A more plausible view is that the stochastic parameter model provides quite a good description of these data over the whole time period.

One might also strengthen this view by noticing that the estimates of ϕ in each of the timeframes are very significantly different from zero. Furthermore, note the statistics for the test of a purely random slope coefficient against the alternative hypothesis that the slope coefficient follows an AR(1) process. This test is a likelihood ratio (LR) test and is described in Section 3.1. These statistics are shown in parentheses under the log-likelihoods in Table 4.3. The statistic for this test, twice the difference between the log-likelihoods of the SPRM (4.7) and the RCM (4.6), is distributed as a chi-square with one degree of freedom. The critical chi-square value for a 5% level test is 3.84. Thus, if the LR test statistic is less than 3.84, then one cannot reject the null hypothesis of a purely random slope coefficient—however, if this test statistic is greater than 3.84, one can conclude that the slope coefficient is more likely to follow an AR(1) process. Note that this test cannot be carried out in the third time frame as the RCM could not be estimated in this time frame. Here also, one can see that the data in the last three time frames strongly support the hypothesis of a slope coefficient that follows an AR(1) process.

Estimation of a stochastic parameter regression model with the intercept and slope both following a first-order autoregressive process. Now that model (4.7), a simple SPRM, seems to be more appropriate than model (4.6), the RCM, the next SPRM that should be estimated is the one in which both α_t and β_t follow an AR(1) process. To cut down on the number of parameters to be estimated, the two AR processes will not be forced into a generalized vector AR format. The next model estimated is

$$I_t = \alpha + \beta_t R_t \qquad [4.8a]$$

$$(\alpha_t - \alpha) = \rho(\alpha_{t-1} - \alpha) + e_t \qquad [4.8b]$$

$$(\beta_t - \beta) = \phi(\beta_{t-1} - \beta) + a_t \qquad [4.8c]$$

where e_t and a_t both have zero means, fixed variances σ_e^2 and σ_a^2, and are both not autocorrelated and are not correlated with each other. In essence, what we have here is a regression model with an autocorrelated intercept and an independently autocorrelated slope. We do not include an error term in the regression equation (4.8a), as this is effectively incorporated in (4.8b), which describes the stochastic behavior of the intercept term. An alternative way of writing (4.8a) – (4.8b) is

$$I_t = \alpha + \beta_t R_t + u_t$$

where

$$u_t = \rho u_{t-1} + e_t$$

Thus, we can also view the model as one with a fixed intercept term, α, and an error term, u_t, which obeys a first-order autoregressive model.

Let a, b, s_e^2, s_a^2, p, and r be the maximum likelihood estimates of the parameters of the SPRM (4.8) α, β, σ_e^2, σ_a^2, ϕ, and ρ. The MLEs of model (4.8) are displayed in Table 4.4. Now that the AR(1) coefficient, ρ, has been included, not much has changed in the estimates of the parameters common to models (4.7) and (4.8), because the estimates of ρ are all insignificantly different from zero. Also, the log-likelihoods of the model (4.8) have not changed much at all from those of the corresponding log-likelihoods of model (4.7). This leads one to believe that model (4.7) is more appropriate than model (4.8) or perhaps any other more general

TABLE 4.4
Maximum Likelihood Estimates of the Stochastic Parameter Regression Model (Equation 4.8)

1953/1 to	*a (SE)*	*b (SE)*	s_e^2 *(SE)*	s_a^2 *(SE)*	*r (SE)*	*p (SE)*	*Log Likelihood*
1965/2	−.6300 (.6471)	.7997 (.2670)	1.3312 (.4245)	.013459 (.035666)	.11171 (.20722)	.83822 (.29469)	−81.770
1968/2	−.5297 (.5764)	.7905 (.1995)	1.3551 (.4059)	.010236 (.030002)	.19443 (.17593)	.78199 (.41892)	−101.042
1971/2	−.6174 (.4965)	.8555 (.1569)	1.3469 (.2874)	.003575 (.009283)	.18933 (.14627)	.88373 (.21257)	−119.257
1974/2	−.8990 (.5416)	1.0802 (.2171)	1.2508 (.2697)	.009426 (.008709)	.11942 (.15202)	.93570 (.06827)	−142.962
1977/2	−.7803 (.5682)	1.0559 (.2105)	1.4400 (.2986)	.009828 (.008721)	.14501 (.14837)	.92965 (.06399)	−169.327
1980/2	−.8659 (.5377)	1.0651 (.1756)	1.5092 (.3175)	.011411 (.008895)	.16983 (.14376)	.91071 (.06445)	−194.950

SPRM. Furthermore, the estimates of any more general SPRM would essentially be those of model (4.7).

Thus the above heuristic argument points to model (4.7) as being the appropriate form of SPRM that relates inflation and the nominal interest rate on 3-month T-bills. This means that α_t seems to be constant through time and that β_t follows an AR(1) process. It really does not matter to the basic argument what ARMA process β_t follows, just that it is not constant over time is sufficient. In other words, the above analysis leads one to believe that the equilibrium real rate of return, or the market's expectation thereof, is constant over time, and that the 3-month T-bill market is not efficient, as β_t is not constant and equal to one. The estimates of equation (4.7b) show that the average β_t, or β, is insignificantly different from one, and so if β_t is thought of as an adjustment factor, the adjustment process is one that brings β_t to unity in the long run. In relation to Fama's orginal study, our analysis suggests that the original choice of time frame for investigation was rather fortuitous, as looking at shorter or longer time frames casts rather more doubt on the conjecture of market efficiency. The results presented here suggest the possibility of a long-run efficiency, but of short-run departures from this long-run path, where the impact of T-bill rates on expected inflation can differ substantially from unity.

4.4 Estimation of the Stochastic Parameters

It is our feeling, based on the information presented above, that the market's adjustment process that brings the rate of return on 3-month T-bills closer to the rate of inflation, is not instantaneous and therefore perfect, but that it is indeed stochastic in a way that is adequately described by equation (4.7b).

It may be of interest to economic historians to know how the adjustment factor, β_t, has actually varied over time. This can easily be estimated with the fixed interval smoothing algorithm, which is described in Section 2.4. This algorithm was used to derive $\beta(t|T)$, the estimates of the β_t, and their sampling variances with data from 1953/1 to 1980/2, the sixth and longest time frame investigated above. These estimates are shown in Table 4.5.

After a look at the $\beta(t|T)$ it is clear that they do indeed change rather smoothly. Note that the estimated standard errors of the $\beta(t|T)$—the square roots of the $P(t|T)$—are quite large. At first one is struck with the fact that the estimated standard errors generally decrease through time. Apparently this phenomenon is a peculiarity of this data set and should not be expected to reoccur in general. Nevertheless, the large estimated standard errors of the $\beta(t|T)$ are to be expected. This happens as we are attempting to estimate T quantities β_t from a data set of T observations—even after estimating the parameters of the SPRM. We are, in a sense, stretching the information in our data set to the limit. Still, we would like to hope that the estimates of the β_t are meaningful in some sense.

To get a clearer impression of how it is estimated that the adjustment factor, β_t, moves through time, the β_t were estimated for each time frame, based on the parameter estimation of the model (4.7) shown in Table 4.3. These six estimated series of β_t are plotted in Figure 4.1. We did this to eliminate any nagging suspicion that the estimated models, and their seemingly conflicting test results, would give rise to vastly different estimates of the β_t. Obviously this is not so, thus strengthening our belief in our conclusion that the market for 3-month T-bills is inefficient.

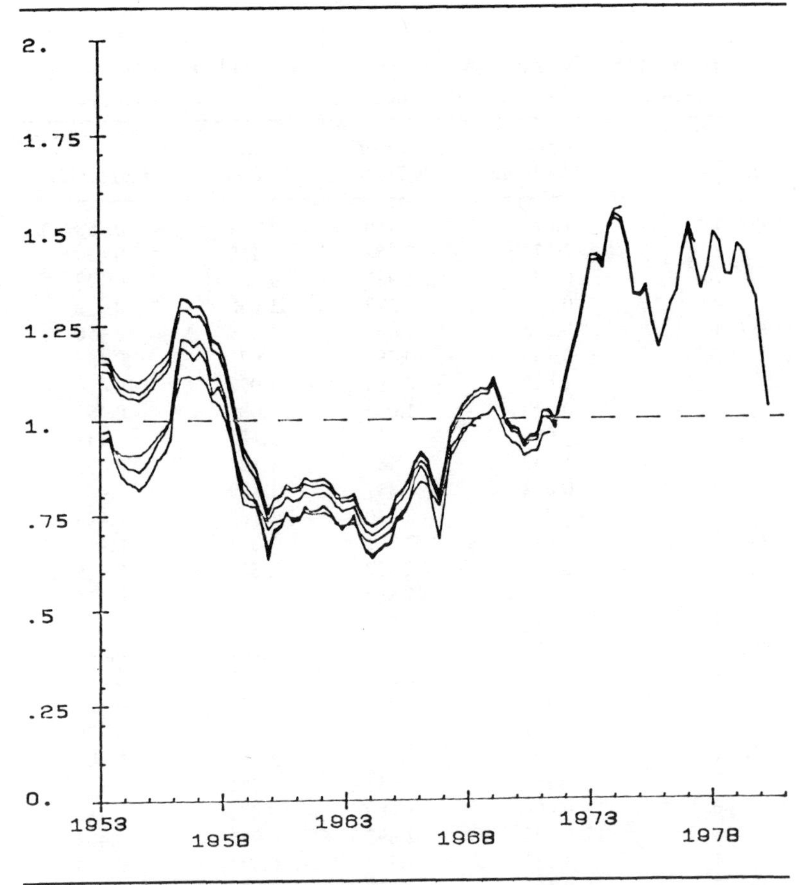

Figure 4.1: Plot of the Estimated β_t for the 6 Time Frames Beginning in the First Quarter of 1953 and Ending in the Second Quarters of 1965, 1968, 1971, 1974, 1977, and 1980, respectively

TABLE 4.5

Data on the Rates on 3-Month Treasury Bills and on Inflation: Estimated Values of β_t and Their Estimated Standard Errors

Date		*Rate on 3-month T-bills*	*Rate of Inflation*	$\beta(t\|T)$	*SE of* $\beta(t\|T)$
1953	1	1.673	1.980	1.151	0.458
	2	3.173	2.153	1.150	0.450
	3	0.492	1.957	1.111	0.449
	4	–0.327	1.473	1.088	0.452
1954	1	–0.333	1.060	1.079	0.456
	2	0.169	0.787	1.077	0.457
	3	–1.822	0.883	1.072	0.456
	4	–0.499	1.017	1.083	0.453
1955	1	0.000	1.227	1.102	0.447
	2	1.333	1.483	1.125	0.439
	3	0.662	1.857	1.139	0.430
	4	–0.830	2.340	1.163	0.420
1956	1	2.983	2.327	1.254	0.413
	2	5.115	2.567	1.320	0.406
	3	2.928	2.583	1.314	0.401
	4	2.417	3.033	1.297	0.394
1957	1	4.011	3.100	1.301	0.391
	2	4.296	3.137	1.275	0.389
	3	1.415	3.353	1.208	0.390
	4	4.071	3.303	1.203	0.396
1958	1	3.104	1.760	1.161	0.409
	2	0.462	0.957	1.081	0.415
	3	0.000	1.680	0.997	0.412
	4	0.000	2.690	0.925	0.403
1959	1	1.231	2.773	0.894	0.395
	2	2.451	3.000	0.870	0.388
	3	2.134	3.540	0.817	0.379
	4	–0.150	4.230	0.753	0.372
1960	1	2.728	3.873	0.792	0.377
	2	0.750	2.993	0.802	0.388
	3	2.407	2.360	0.830	0.397
	4	0.148	2.307	0.819	0.403
1961	1	0.148	2.350	0.824	0.405
	2	2.091	2.303	0.846	0.407
	3	0.445	2.303	0.838	0.406
	4	0.890	2.460	0.840	0.404
1962	1	1.776	2.723	0.844	0.400
	2	1.618	2.713	0.832	0.398
	3	0.885	2.840	0.807	0.395
	4	0.584	2.813	0.792	0.393

(continued)

TABLE 4.5 (Continued)

Date		Rate on 3-month T-bills	Rate of Inflation	$\beta(t\|T)$	SE of $\beta(t\|T)$
1963	1	1.022	2.907	0.795	0.390
	2	2.918	2.937	0.803	0.387
	3	1.012	3.293	0.760	0.382
	4	1.014	3.497	0.731	0.378
1964	1	0.864	3.530	0.719	0.376
	2	1.436	3.477	0.730	0.375
	3	1.577	3.497	0.742	0.373
	4	0.710	3.683	0.751	0.370
1965	1	2.849	3.890	0.800	0.367
	2	1.837	3.873	0.815	0.366
	3	1.689	3.867	0.841	0.363
	4	3.221	4.167	0.889	0.359
1966	1	4.038	4.610	0.913	0.353
	2	3.853	4.587	0.895	0.350
	3	2.998	5.043	0.840	0.346
	4	0.812	5.210	0.796	0.346
1967	1	2.702	4.513	0.881	0.352
	2	4.290	3.660	0.978	0.358
	3	3.321	4.300	1.008	0.355
	4	4.209	4.753	1.039	0.349
1968	1	4.427	5.050	1.057	0.343
	2	5.028	5.520	1.071	0.338
	3	4.832	5.197	1.080	0.337
	4	4.522	5.587	1.077	0.333
1969	1	6.966	6.093	1.107	0.327
	2	5.740	6.197	1.058	0.323
	3	5.539	7.023	0.998	0.316
	4	5.941	7.353	0.975	0.312
1970	1	6.437	7.210	0.971	0.314
	2	4.496	6.677	0.935	0.319
	3	5.240	6.330	0.952	0.326
	4	3.036	5.353	0.955	0.339
1971	1	4.574	3.840	1.020	0.353
	2	3.976	4.250	1.017	0.354
	3	2.186	5.010	0.984	0.352
	4	3.149	4.230	1.047	0.358
1972	1	3.235	3.437	1.123	0.364
	2	3.639	3.770	1.186	0.362
	3	3.603	4.220	1.245	0.355
	4	5.559	4.863	1.337	0.345
1973	1	8.811	5.700	1.430	0.332
	2	8.807	6.603	1.427	0.319
	3	9.331	8.323	1.402	0.305
	4	11.233	7.500	1.504	0.307

(continued)

TABLE 4.5 (Continued)

Date		Rate on 3-month T-bills	Rate of Inflation	$\beta(t\|T)$	SE of $\beta(t\|T)$
1974	1	11.316	7.617	1.536	0.306
	2	12.187	8.153	1.525	0.303
	3	11.667	8.190	1.456	0.303
	4	7.267	7.360	1.329	0.312
1975	1	6.282	5.750	1.327	0.326
	2	8.529	5.393	1.351	0.331
	3	6.384	6.330	1.250	0.328
	4	3.870	5.627	1.188	0.334
1976	1	4.943	4.917	1.239	0.341
	2	6.387	5.157	1.302	0.341
	3	4.501	5.150	1.332	0.343
	4	7.049	4.673	1.442	0.346
1977	1	8.596	4.630	1.492	0.347
	2	5.834	4.840	1.413	0.343
	3	4.436	5.497	1.341	0.335
	4	6.760	6.110	1.387	0.328
1978	1	10.391	6.393	1.487	0.323
	2	9.454	6.477	1.463	0.319
	3	8.088	7.313	1.377	0.310
	4	9.958	8.570	1.375	0.298
1979	1	13.719	9.383	1.457	0.291
	2	13.210	9.377	1.438	0.289
	3	11.698	9.673	1.354	0.285
	4	15.568	11.843	1.318	0.271
1980	1	14.447	13.353	1.164	0.264
	2	7.571	9.617	1.030	0.302

REFERENCES

ANDERSON, B.D.O. and J. B. MOORE (1979) Optimal Filtering. Englewood Cliffs, NJ: Prentice-Hall.

ANDERSON, T. W. (1958) An Introduction to Multivariate Statistical Analysis, New York: John Wiley.

ALEXANDER, G. J. and G. BENSEN (1982) "More on beta as a random coefficient." Journal of Financial and Quantitative Analysis 17 (1): 27-36.

ANSLEY, C. F. and R. KOHN (1982) "A geometrical derivation of the fixed interval smoothing algorithm." Biometrika 69: 486-487.

BLUME, M. (1975) "Betas and their regression tendencies." Journal of Finance 30: 785-795.

BOS, T. (1982) "Exact maximum likelihood estimation of the Kalman filter model." Ph.D. dissertation, Department of Economics, University of Illinois.

———and P. NEWBOLD (1984) "An empirical investigation of the possibility of stochastic systematic risk in the market model." Journal of Business 57: 35-41.

BOX, G.E.P. and G. M. JENKINS (1970) Time Series Analysis, Forecasting and Control. San Francisco: Holden Day.

BREUSCH, T. S. and A. R. PAGAN (1980) "The Lagrange multiplier test and its applications to model specification in econometrics." Review of Economic Studies 47: 239-254.

———(1979) "A simple test for heteroscedasticity and random coefficient variation." Econometrica 47: 1287-1294.

CHANT, D. (1974) "On asymptotic tests of composite hypotheses in non-standard conditions." Biometrika 61 (2): 291-298.

COOLEY, T. F. and E. C. PRESCOTT (1973) "Varying parameter regression: a theory and some applications." Annals of Economic and Social Measurement No. 2/4, 463-474.

DAVIES, R. B. (1977) "Hypothesis testing when a nuisance parameter is present only under the alternative." Biometrika 64 (2): 247-254.

ENGLE, R. and M. WATSON (1981) "A one-factor multivariate time series model of metropolitan wage rates." Journal of the American Statistical Association 76: 774-781.

FABOZZI, F. J. and J. C. FRANCIS (1978) "Beta as a random coefficient." Journal of Financial and Quantitative Analysis 13: 101-116.

———and C. F. LEE (1982) "Specification error, random coefficient, and risk return relationship." Quarterly Review of Economics and Business 22 (1): 23-31.

FAMA, E. F. (1977) "Interest rates and inflation: the message in the entrails." American Economic Review 67: 487-496.

———(1975) "Short-term interest rates as predictors of inflation." American Economic Review 65: 269-282.

———(1970) "Efficient capital markets: a review of theory and empirical work." Journal of Finance 25: 383-417.

———(1968) "Risk, return and equilibrium: some clarifying comments." Journal of Finance 23: 29-40.

GRANGER, C.W.J. and P. NEWBOLD (1977) Forecasting Economic Time Series, New York: Academic.

GUPTA, N. and R. K. MEHRA (1974) "Computational aspects of maximum likelihood estimation and reduction in sensitivity function calculations." IEEE Transactions on Automatic Control AC-19(6): 774-783.

HAVENNER, A. and P.A.V.B. SWAMY (1981) "A random coefficient approach to seasonal adjustment of economic time series." Journal of Econometrics 15: 177-209.

HILDRETH, C. and J. P. HOUCK (1968) "Some estimators for a linear model with random coefficients." Journal of the American Statistical Association 63: 584-595.

JACOB, N. L. (1971) "The measurement of systematic risk for securities and portfolios: some empirical results." Journal of Financial and Quantitative Analysis 6: 815-834.

JENKINS, G. M. and A. S. ALAVI (1981) "Some aspects of modelling and forecasting multivariate time series." Journal of Time Series Analysis 2: 1-47.

KALMAN, R. E. (1960) "A new approach to linear filtering and prediction problems." Journal of Basic Engineering 82: 34-45.

———and R. S. BUCY (1961) "New results in linear filtering and prediction theory." ASME Transactions, Series D, Journal of Basic Engineering 83: 95-107.

LEE, C. F. and S. N. CHEN (1980) "A random coefficient model for reexamining risk-decomposition method and risk-return relationship test." Quarterly Review of Economics and Business 20(4): 58-69.

LINTNER, J. (1965a) "The valuation of risk assets and the selection of risky investments in stock portfolios and capital budgets." Review of Economics and Statistics 48(1): 13-37.

———(1965b) "Security prices, risk, and maximal gains from diversification." Journal of Finance 20(4): 587-615.

NELSON, C.R. (1973) Applied Time Series Analysis for Managerial Forecasting. San Francisco: Holden Day.

OHLSON, J. and B. ROSENBERG (1982) "Systematic risk of the CRSP equal-weighted common stock index: a history estimated by stochastic parameter regression." Journal of Business 55: 121-145.

PAGAN, A.R. (1980) "Some identification and estimation results for regression models with stochastically varying coefficients." Journal of Econometrics 13: 341-363.

RAO, C.R. (1948) "Large sample tests of statistical hypotheses concerning several parameters with application to problems of estimation." Proceedings of the Cambridge Philosophical Society 44: 50-57.

ROSENBERG, B. (1973) "Linear regression with randomly dispersed parameters reexamined." Biometrika 60(1) 65-72.

———(1972) "The estimation of stationary stochastic regression parameters reexamined." Journal of the American Statistical Association 67: 650-654.

———and J. GUY (1976a) "Prediction of beta from investment fundamentals, part one: prediction criteria." Financial Analysts Journal (May-June): 60-72.

———(1976b) "Prediction of beta from investment fundamentals, part two: alternate prediction methods." Financial Analysts Journal (July-August): 62-70.

SAGE, A. P. and J. L. MELSA (1971) Estimation Theory with Applications to Communications and Control. New York: McGraw-Hill.

SCHWEPPE, F. C. (1965) "Evaluation of likelihood functions for Gaussian signals." IEEE Transactions on Information Theory IT-11: 61-70.

SHARPE, W. F. (1964) "Capital asset prices: a theory of market equilibrium under conditions of risk." Journal of Finance 19(3): 425-442.

SILVEY, D. S. (1959) "The Lagrangean multiplier test." Annals of Mathematical Statistics 30: 389-407.

SUNDER, S. (1980) "Stationarity of market risk: random coefficients tests for individual stocks." Journal of Finance 35(4): 883-896.

SWAMY, P.A.V.B. and P. A. TINSLEY (1980) "Linear prediction and estimation methods for regression models with stationary stochastic coefficients." Journal of Econometrics 12: 103-142.

TIAO, G.C. and G.E.P. BOX (1981) "Modeling Multiple Time Series with Applications," Journal of the American Statistical Association 76: 802-816.

WATSON, M. (1980) "Applications of Kalman filter models in econometrics." Ph.D. dissertation, University of California, San Diego.

PAUL NEWBOLD is Professor of Economics at the University of Illinois at Urbana-Champaign. His Ph.D. is in Statistics from the University of Wisconsin. He is co-author, with C.W.J. Granger, of Forecasting Economic Time Series *(Academic Press, 1977).*

THEODORE BOS is Assistant Professor in Quantitative Methods in the Department of Economics, University of Alabama at Birmingham. His current work deals with various issues in econometrics and time-series analysis.

SAGE JOURNALS

The Publishers of Professional Social Science
2111 West Hillcrest Drive, Newbury Park, California 91320

rders from the U.K., Europe, the Middle East, and Africa should be sent to: **Sage Publications, Ltd. 28 Banner Street London EC1Y 8QE ENGLAND**

Orders from India and South Asia should be sent to: **Sage Publications India Private Limited P.O. Box 4215 New Delhi 110 048 INDIA**

These journals are available from Sage Periodicals Press. See following pages for journals available from Sage, London.

TRACTS IN SOCIAL GERONTOLOGY
provides abstracts and bibliographies of major articles, books, rts, and other materials on all aspects of gerontology: in- ing demography, economics, family relations, government cy, health, institutional care, physiology, psychiatric dysfunc- s, psychology, societal attitudes, work and retirement.
rterly: March, June, Sept., Dec.
ly rates: Inst. $98 / Ind. $48
ST ISSUE: MARCH 1990 / ISSN: 1047-4862

INISTRATION & SOCIETY
or: Gary L. Wamsley,
inia Polytechnic Institute and State Univ.
deals with administration, bureaucracy, public organization, public policy — and the impact these have on politics and ety.
rterly: May, Aug., Nov., Feb.
ly rates: Inst. $99 / Ind. $39 / ISSN: 0095-3997

ILIA: Journal of Women and Social Work
or-in-Chief: Betty Sancier, *Univ. of Wisconsin-Milwaukee*
s a publication for and about women social workers and their ts. Its intent is to bring insight and knowledge to the field ocial work from a feminist perspective and to provide the arch and tools necessary to make significant changes and ovements in the delivery of social services.
rterly: Feb., May, Aug., Nov.
ly Rates: Inst. $64 / Ind. $30 / ISSN: 0886-1099

ERICAN BEHAVIORAL SCIENTIST
ocuses, in theme-organized issues prepared under guest ors, on emerging cross-disciplinary interests, research, and lems in the social sciences.
onthly: Sept., Nov., Jan., Mar., May, July
ly rates: Inst. $108 / Ind. $36 / ISSN: 0002-7642

ERICAN POLITICS QUARTERLY
or: Susan Welch, *Univ. of Nebraska*
promotes basic research in all areas of American political avior — including urban, state, and national policies, as well ressing social problems requiring political solutions.
rterly: Jan., April, July, Oct.
ly rates: Inst. $98 / Ind. $34 / ISSN: 0044-7803

ANNALS
he American Academy of Political and Social Science
or: Richard D. Lambert
ociate Editor: Alan W. Heston
Since 1891, The Annals has served as the preeminent forum he interdisciplinary discussion of single problems and policy es affecting America and the world community.
onthly: Jan., March, May, July, Sept., Nov.
ly rates: Inst. $72 (p) / $89 (c) Ind. $32 (p) / $45 (c)
N: 0002-7162

AVIOR MODIFICATION
ors: Michel Hersen, *Western Psychiatric Inst. & Clinic*
n S. Bellack, *Medical College of Pennsylvania at EPPI*
describes (in detail for replication purposes) assessment and ification techniques for problems in psychiatric, clinical, cational, and rehabilitational settings.
rterly: Jan., April, July, Oct.
ly rates: Inst. $100 / Ind. $38 / ISSN: 0145-4455

NA REPORT: A Journal of East Asian Studies
or: C.R.M. Rao,
ntre for the Study of Developing Societies, Delhi
encourages the increased understanding of contemporary na and its East Asian neighbors, their cultures and ways of elopment, and their impact on India and other South Asian ntries.
rterly: Feb., May, Aug., Nov.
ly rates: Inst. $59 / Ind. $30 / ISSN: 0009-4455

COMMUNICATION ABSTRACTS
Editor: Thomas F. Gordon, *Temple Univ.*
. . . provides coverage of recent literature in all areas of communication studies (both mass and interpersonal). Includes expanded coverage of new communications technologies.
Bimonthly: Feb., April, June, Aug., Oct., Dec.
Yearly rates: Inst. $270 / Ind. $90 / ISSN: 0162-2811

COMMUNICATION RESEARCH
Editor: Peter R. Monge, *Annenberg School, USC*
. . . provides an interdisciplinary forum for scholars and professionals to present new research in communication. Encourages rigorous studies of mass (and interpersonal) communication.
Bimonthly: Feb., April, June, Aug., Oct., Dec.
Yearly rates: Inst. $132 / Ind. $42 / ISSN: 0093-6502

COMPARATIVE POLITICAL STUDIES
Editor: James A. Caporaso, *Univ. of Washington*
. . . publishes theoretical and empirical research articles by scholars engaged in cross-national study, and includes research notes and review essays.
Quarterly: April, July, Oct, Jan.
Yearly rates: Inst. $96 / Ind. $32 / ISSN: 0010-4140

CONTRIBUTIONS TO INDIAN SOCIOLOGY
Editor: T. N. Madan, *Institute of Economic Growth, Delhi*
. . . a distinguished international forum for research on Indian and South Asian societies.
Biannually: May and November
Yearly rates: Inst. $59 / Ind. $27 / ISSN: 0069-9667

The COUNSELING PSYCHOLOGIST
Journal of Counseling Psychology of the American Psychological Association (Div. 17)
Editor: Bruce R. Fretz, *Univ. of Maryland, College Park*
Editor-elect: Gerald Stone, *Univ. of Iowa*
. . . presents timely coverage — especially in new or developing areas of practice and research — of topics of immediate interest to counseling psychologists. Defines the field and communicates that identity to the profession as well as to those in other disciplines.
Quarterly: Jan., April, July, Oct.
Yearly rates: Inst. $90 / Ind. $32 / ISSN: 0011-0000

CRIME & DELINQUENCY
Published in cooperation with the National Council on Crime and Delinquency
Editor: Don C. Gibbons, *Portland State Univ.*
. . . addresses specific policy or program implications or issues — social, political, and economic — of great topical interest to the professional with direct involvement in criminal justice.
Quarterly: Jan., April, July, Oct.
Yearly rates: Inst. $96 / Ind. $36 / ISSN: 0011-1287

CRIMINAL JUSTICE AND BEHAVIOR
Official Publication of the AACP
Published in affiliation with the ACA
Editor: Allen K. Hess, *Auburn Univ.*
. . . provides a means of communication among mental health professionals, behavioral scientists, researchers, and practitioners in the area of criminal justice.
Quarterly: March, June, Sept., Dec.
Yearly rates: Inst. $95 / Ind. $36 / ISSN: 0093-8548

Sage Periodicals Press — A Division of SAGE Publications, Inc.
Newbury Park • London • New Delhi

ECONOMIC DEVELOPMENT QUARTERLY
The Journal of American Economic Revitalization
Editors: Richard D. Bingham, *Cleveland State*
Sammis B. White, *Univ. of Wisconsin-Milwaukee*
& Gail Garfield Schwartz, *NY State Public Serv. Commission*
. . .disseminates information on the latest research, programs, policies, and trends in the field of economic development. EDQ is unique in its concern for all areas of development — large cities, small towns, rural areas, and overseas trade and expansion.
Quarterly: Feb., May, Aug., Nov.
Yearly rates: Inst. $90 / Ind. $36 / ISSN: 0891-2424

EDUCATION and URBAN SOCIETY
. . .provides, through theme-organized issues prepared under guest editors, a forum for social scientific research on education as a social institution within urban environments, the politics of education, and educational institutions and processes as agents of social change.
Quarterly: Nov., Feb., May, Aug.
Yearly rates: Inst. $90 / Ind. $34 / ISSN: 0013-1245

EDUCATIONAL ADMINISTRATION ABSTRACTS
. . .provides abstracts drawn from more than 140 professional Journals relating to educational administration.
Quarterly: Jan., April, July, Oct.
Yearly rates: Inst. $188 / Ind. $66 / ISSN: 0013-1601

EDUCATIONAL ADMINISTRATION QUARTERLY
Published in cooperation with the
University Council for Educational Administration
Editor: Steven T. Bossert, *Univ. of Utah*
. . .seeks to stimulate critical thought and to disseminate the latest knowledge about research and practice in educational administration.
Quarterly: Feb., May, Aug., Nov.
Yearly rates: Inst. $90 / Ind. $36 / ISSN: 0013-161X

ENVIRONMENT AND BEHAVIOR
Published in cooperation with the
Environmental Design Research Association (edra)
Editor: Robert B. Bechtel, *Univ. of Arizona*
. . .reports rigorous experimental and theoretical work on the study, design, and control of the physical environment and its interaction with human behavioral systems.
Bimonthly: Jan., Mar., May, July, Sept., Nov.
Yearly rates: Inst. $120 / Ind. $48 / ISSN: 0013-9165

EVALUATION & THE HEALTH PROFESSIONS
Editor: R. Barker Bausell, *Univ. of Maryland*
. . .provides a forum for all health professionals interested or engaged in the development, implementation, and evaluation of health programs.
Quarterly: March, June, Sept., Dec.
Yearly rates: Inst. $96 / Ind. $36 / ISSN: 0163-2787

EVALUATION REVIEW
A Journal of Applied Social Research
Editors: Richard A. Berk & Howard E. Freeman,
both at Univ. of California, Los Angeles
. . .a forum for researchers, planners, and policymakers engaged in the development, implementation, and utilization of evaluation studies. Reflects a wide range of methodological and conceptual approaches to evaluation and its many applications.
Bimonthly: Feb., April, June, Aug., Oct., Dec.
Yearly rates: Inst. $120 / Ind. $45 / ISSN: 0193-841X

GENDER & SOCIETY
Official Publication of Sociologists for Women in Society
Editor: Judith Lorber,
Graduate School and Brooklyn College, CUNY
. . .focuses on the social and structural study of gender as a basic principle of the social order and as a primary social category. Emphasizing theory and research, G&S aims to advance both the study of gender and feminist scholarship.
Quarterly: March, June, Sept., Dec.
Yearly rates: Inst. $84 / Ind. $32 / ISSN: 0891-2432

GROUP & ORGANIZATION STUDIES
An International Journal
Editor: Michael J. Kavanagh, *SUNY, Albany*
. . .bridges the gap between research and practice for psychologists, group facilitators, educators, and consultants who are involved in the broad field of human relations training.
Quarterly: March, June, Sept., Dec.
Yearly rates: Inst. $100 / Ind. $42 / ISSN: 0364-1082

HISPANIC JOURNAL OF BEHAVIORAL SCIENCES
Editor: Amado M. Padilla, *Stanford University*
. . .publishes research articles, case histories, critical review scholarly notes that are of theoretical interest or deal methodological issues related to Hispanic populations.
Quarterly: Feb., May, Aug., Nov.
Yearly rates: Inst. $60 / Ind. $30 / ISSN: 0739-9863

HUMAN COMMUNICATION RESEARCH
Editor: James Bradac, *Univ. of Calif, Santa Barbara*
. . .publishes important research and high-quality reports contribute to the expanding body of knowledge about hu communication.
Quarterly: Sept., Dec., March, June
Yearly rates: Inst. $96 / Ind. $36 / ISSN: 0360-3989

HUMAN RESOURCES ABSTRACTS
. . .contains abstracts of the most important recent literatu the professional who needs easy reference to current and ch ing ideas in the diverse area of manpower and human reso development, and related social/governmental policy ques
Quarterly: March, June, Sept., Dec.
Yearly rates: Inst. $188 / Ind. $66 / ISSN: 0099-2453

INDIAN ECONOMIC AND SOCIAL HISTORY REVIEW
Editor: Dharma M. Kumar, *Delhi School of Economics*
. . .focuses on the history, economy, and society of India South Asia, and includes comparative studies of development.
Quarterly: March, June, Sept., Dec.
Yearly rates: Inst. $70 / Ind. $35 / ISSN: 0019-4646

THE INDIAN JOURNAL OF SOCIAL SCIENCE
Sponsored by the Indian Council of Social Science Rese
Editor: Sukhamoy Chakravarty, *Univ. of Delhi*
. . .promotes scientific discussion on the diverse concer social science research — the problems of development social change, the interface between science, society, culture technology, and a comprehension of future patterns of dev ment as they relate to the developing countries.
Quarterly: March, June, Sept., Dec.
Yearly Rates: Inst. $59 / Ind. $35

INTERNATIONAL STUDIES
Editor: Anirudha Gupta, *School of International Studies, Jawaharlal Nehru Univ., New Delhi*
. . .the most outstanding Indian research journal in the fie international affairs and area studies.
Quarterly: Jan., April, July, Oct.
Yearly rates: Inst. $70 / Ind. $35 / ISSN: 0020-8817

JOURNAL OF ADOLESCENT RESEARCH
Editor: E. Ellen Thornburg, *Tucson, Arizona*
. . .provides professionals with the most current and releva formation on ways in which individuals ages 10-20 dev behave, and are influenced by societal and cultural perspec
Quarterly: Jan., April, July, Oct.
Yearly Rates: Inst. $78 / Ind. $36 / ISSN: 0743-5584

JOURNAL OF AGING AND HEALTH
Editor: Kyriakos S. Markides,
Univ. of Texas Medical Branch, Galveston
. . .deals with social and behavioral factors related to aging health, emphasizing health and quality of life.
Quarterly: Feb., May, Aug., Nov.
Yearly Rates: Inst. $78 / Ind. $36 / ISSN: 0898-2643

JOURNAL OF APPLIED GERONTOLOGY
The Official Journal of the Southern Gerontological So
Editor: Miles Simpson, *North Carolina Central Univ.*
. . .strives to consistently publish articles in all subdisciplin aging whose findings, conclusions, or suggestions have clea sometimes immediate applicability to the problems encount by older persons.
Quarterly: March, June, Sept., Dec.
Yearly rates: Inst. $88 / Ind. $38 / ISSN: 0733-4648

JOURNAL OF BLACK STUDIES
Editor: Molefi Kete Asante, *Temple Univ.*
. . .sustains full analytical discussion of economic, poli sociological, historical, literary, and philosophical issues re to persons of African descent.
Quarterly: Sept., Dec., March, June
Yearly rates: Inst. $92 / Ind. $34 / ISSN: 0021-9347

JOURNAL OF CONFLICT RESOLUTION
Journal of The Peace Science Society (International)
Editor: Bruce M. Russett, *Yale University*
. . .draws from interdisciplinary sources in its focus on the ana of causes, prevention, and solution of international, domestic interpersonal conflicts.
Quarterly: March, June, Sept., Dec.
Yearly rates: Inst. $116 / Ind. $40 / ISSN: 0022-0027

RNAL OF CONTEMPORARY ETHNOGRAPHY
nerly Urban Life)
ors: Patricia Adler, *Univ of Colorado, Boulder*
ter Adler, *Univ. of Denver*
ne first journal dedicated to ethnography and qualitative
arch in general. Advances sociological knowledge through
sive, in-depth studies of human behavior in natural settings.
terly: April, July, Oct., Jan.
y rates: Inst. $105 / Ind. $34 / ISSN: 0891-2416

RNAL OF CROSS-CULTURAL PSYCHOLOGY
lished for the Center for Cross-Cultural Research,
tern Washington University
or: Juris G. Draguns, *Pennsylvania State Univ.*
or Editor: Walter J. Lonner, *Western Washington U.*
presents behavioral and social research concentrating on
hological phenomena as differentially conditioned by culture,
on the individual as a member of the cultural group.
rterly: March, June, Sept., Dec.
ly rates: Inst. $94 / Ind. $35 / ISSN: 0022-0221

RNAL OF EARLY ADOLESCENCE
or: E. Ellen Thornburg, *Tucson, Arizona*
provides a well-balanced, interdisciplinary, international
pective on early adolescent development (age 10 through 14
s) and the factors affecting it.
rterly: Feb., May, Aug., Nov.
ly rates: Inst. $68 / Ind. $32 / ISSN: 0272-4316

RNAL OF FAMILY ISSUES
nsored by the National Council on Family Relations
or: Patricia A. Voydanoff, *Univ. of Dayton*
devoted to contemporary social issues and social problems
ed to marriage and family life, and to theoretical and profes-
al issues of current interest to those who work with and study
lies.
rterly: March, June, Sept., Dec.
ly rates: Inst. $95 / Ind. $35 / ISSN: 0192-513X

JRNAL OF FAMILY PSYCHOLOGY
rnal of the Division of Family Psychology of the
erican Psychological Association (Div. 43)
or: Howard A. Liddle, *Temple Univ.*
enhances theory, research, and clinical practice in family
chology and deals with: family and marital theory and con-
ts; research and evaluation; therapeutic frame works and
hods; training and supervision; policies and legal matters con-
ning the family and marriage.
rterly: Sept., Dec., March, June
ly rates: Inst. $80 / Ind. $36 / ISSN: 0893-3200

RNAL OF HUMANISTIC PSYCHOLOGY
lished in cooperation with the
ociation for Humanistic Psychology
or: Thomas Greening, *Psychological Service Associates*
provides an interdisciplinary forum for contributions and con-
ersies in humanistic psychology as applied to personal growth,
personal encounter, social problems, and philosophical
es.
rterly: Jan, April, July, Oct.
ly rates: Inst. $90 / Ind. $34 / ISSN: 0022-1678

RNAL OF INTERPERSONAL VIOLENCE
cerned with the Study and Treatment of Victims and
petrators of Physical and Sexual Violence
or: Jon R. Conte, *Univ. of Chicago*
provides a forum for discussion of the concerns and activities
rofessionals and researchers working in domestic violence,
d sexual abuse, rape and sexual assault, physical child abuse,
violent crime.
rterly: March, June, Sept., Dec.
ly rates: Inst. $80 / Ind. $35 / ISSN: 0886-2605

RNAL OF MENTAL HEALTH COUNSELING
cial Publication of the
erican Mental Health Counselors Association
or: Lawrence Gerstein, *Ball State University*
disseminates pertinent theory, therapeutic applications, and
arch related to mental health counseling.
rterly: Jan., Apr., July, Oct.
ly rates: Inst. $60 / Ind. $26 / ISSN: 0193-1830

JOURNAL OF URBAN HISTORY
Editor: Blaine A. Brownell, *Univ. of Alabama, Birmingham*
. . . studies the history of cities and urban societies in all periods of human history and in all geographical areas of the world.
Quarterly: Nov., Feb., May, Aug.
Yearly rates: Inst. $98 / Ind. $34 / ISSN: 0096-1442

JOURNAL OF RESEARCH IN CRIME AND DELINQUENCY
Published in Cooperation with the
National Council on Crime and Delinquency
Editor: Vincent O'Leary, *SUNY Albany*
. . . reports on original research in crime and delinquency, new theory, and the critical analyses of theories and concepts especially pertinent to research development in this field.
Quarterly: Feb., May, Aug., Nov.
Yearly rates: Inst. $95 / Ind $36 / ISSN: 0022-4278

KNOWLEDGE:
Creation, Diffusion, Utilization
Editor: Robert Rich, *Univ. of Illinois*
. . . provides a forum for researchers, policymakers, R&D managers, and practitioners engaged in the process of knowledge development which includes the processes of creation, diffusion, and utilization.
Quarterly: Sept., Dec., March, June
Yearly rates: Inst. $95 / Ind. $38 / ISSN: 0164-0259

LATIN AMERICAN PERSPECTIVES
A Journal on Capitalism and Socialism
Managing Editor: Ronald H. Chilcote,
Univ. of California, Riverside
. . . discusses and debates critical issues relating to capitalism, imperialism, and socialism as they affect individuals, societies, and nations throughout the Americas.
Quarterly: Jan., April, July, Oct.
Yearly rates: Inst. $95 / Ind. $32 / ISSN: 0094-582X

MANAGEMENT COMMUNICATION QUARTERLY
An International Journal
Editors: Paul C. Feingold, *USC*
Christine Kelly, *New York Univ.*
Larry R. Smeltzer, *Arizona State Univ.*
JoAnne Yates, *MIT*
. . . brings together communication research from a wide variety of fields, with a focus on managerial and organizational effectiveness. Includes book reviews and notes from professionals in the field.
Quarterly: Aug., Nov., Feb., May
Yearly rates: Inst. $85 / Ind. $32 / ISSN: 0893-3189

MODERN CHINA
An International Quarterly of History and Social Science
Editor: Philip C. C. Huang, *Univ. of California, Los Angeles*
. . . encourages a new interdisciplinary scholarship and dialogue on China's ongoing revolutionary experience.
Quarterly: Jan., April, July, Oct.
Yearly rates: Inst. $98 / Ind. $39 / ISSN: 0097-7004

PEACE & CHANGE
Sponsored by the Council on Peace Research in History (CPRH) & the Consortium on Peace Research, Education and Development (COPRED)
Editors: Robert D. Schulzinger & Paul Wehr,
University of Colorado-Boulder
. . . publishes scholarly and interpretive articles related to the achieving of a peaceful, just, and humane society. It seeks to transcend national, disciplinary, and occupational boundaries and to build bridges between peace research, education, and action.
Quarterly: Jan., Apr., July, Oct.
Yearly rates: Inst. $60 / Ind. $30 / ISSN: 0149-0508

PERSONALITY AND SOCIAL PSYCHOLOGY BULLETIN
Journal of the Society for Personality and Social Psychology
Editor: Richard E. Petty, *Ohio State Univ.*
. . . publishes theoretical articles and empirical reports of research in all areas of personality and social psychology.
Quarterly: March, June, Sept., Dec.
Yearly rates: Inst. $120 / Ind. $44 / ISSN: 0146-1672

PERSON-CENTERED REVIEW
An International Journal of Research, Theory, and Application
Editor: David J. Cain
. . . is devoted to the continued development of person-centered theory, research, and application in the fields of psychotherapy, education, supervision and training, and human development in various group and organizational settings.
Quarterly: Feb., May, Aug., Nov.
Yearly rates: Inst. $80 / Ind. $35 / ISSN: 0883-2293

PHILOSOPHY OF THE SOCIAL SCIENCES **New**
Editors: John O'Neill, I.C. Jarvie,
J.N. Hattiangadi, *York University, Toronto*
. . . publishes articles, discussions, symposia, literature surveys, and more of interest both to philosophers concerned with the social sciences and to social scientists concerned with the philosophical foundations of their subject.
Quarterly: March, June, Sept., Dec.
Yearly rates: Inst. $70 / Ind. $35 / ISSN: 0048-3931

POLITICAL THEORY
An International Journal of Political Philosophy
Editor: Tracy B. Strong, *Univ. of Calif., San Diego*
. . . provides a forum for the diverse orientations in the study of political ideas, including the history of political thought, modern theory, conceptual analysis, and polemic argumentation.
Quarterly: Feb., May, Aug., Nov.
Yearly rates: Inst. $98 / Ind. $35 / ISSN: 0090-5917

PSYCHOLOGY AND DEVELOPING SOCIETIES
A Journal Published by the Centre of Advanced Study in Psychology, Univ. of Allahabad, India
Chief Editor: Durganand Sinha, *National Fellow, Indian Council for Social Science Research, New Delhi,*
. . . provides a forum for psychologists from different parts of the world who are concerned with problems of developing societies. The journal will publish theoretical, empirical, and review papers which help to further understanding of the problems of these societies.
FIRST ISSUE: March, 1989 / Bi-Annual: March, Sept.
Yearly rates: Inst. $49 / Ind. $24

PUBLIC FINANCE QUARTERLY
Editor: J. Ronnie Davis, *Univ. of New Orleans—Lakefront*
. . . studies the theory, policy, and institutions related to the allocation, distribution, and stabilization functions within the public sector of the economy.
Quarterly: Jan., April, July, Oct.
Yearly rates: Inst. $115 / Ind. $44 / ISSN: 0048-5853

RATIONALITY AND SOCIETY
Editor: James S. Coleman, *University of Chicago*
. . . focuses on the growing contributions of rational-action based theory, and the questions and controversies surrounding this growth. The journal publishes work in social theory and social research based on the rational-action paradigm, as well as work challenging this paradigm.
First Issue, July 1989
2 issues in 1989: July, Oct. Quarterly in 1990
Rates: Inst. $141 / Ind. $57 (Vol 1&2-6 issues) / ISSN: 1043-4631

RESEARCH ON AGING
A Quarterly of Social Gerontology and Adult Development
Editors: Rhonda J.V. Montgomery, *Inst. of Gerontology, Wayne State Univ.*
& Edgar F. Borgatta, *Inst. on Aging, Univ. of Washington,*
. . . a journal of interdisciplinary research on current issues, methodological and research problems in the study of the aged.
Quarterly: March, June, Sept., Dec.
Yearly rates: Inst. $98 / Ind. $35 / ISSN: 0164-0275

SAGE FAMILY STUDIES ABSTRACTS
. . . abstracts major articles, reports, books and other materials on policy, theory, and research relating to the family, traditional and alternative lifestyles, therapy and counseling.
Quarterly: Feb., May, Aug., Nov.
Yearly rates: Inst. $188 / Ind. $66 / ISSN: 0164-0283

SAGE PUBLIC ADMINISTRATION ABSTRACTS
. . . publishes cross-indexed abstracts covering recent literature (plus related citations) on all aspects of public administration. Entries are drawn from books, articles, pamphlets, government publications, significant speeches, legislative research studies, and other fugitive material.
Quarterly: April, July, Oct., Jan.
Yearly rates: Inst. $188 / Ind. $66 / ISSN: 0094-6958

SAGE URBAN STUDIES ABSTRACTS
. . . publishes cross-indexed abstracts of important recent literature (plus related citations) on all aspects of urban studies: government and administration, policy, transportation, spatial analysis, planning, social analysis, community studies, education, finance and economics, law, management, environment, and comparative urban analysis.
Quarterly: Feb., May, Aug., Nov.
Yearly rates: Inst. $188 / Ind. $66 / ISSN: 0090-5747

SCIENCE, TECHNOLOGY, & HUMAN VALUES
Sponsored by the Society for Social Studies of Science (4S)
Editor: Susan E. Cozzens, *Rensselaer Polytechnic Institute*
. . . contains research and commentary on the development and dynamics of science and technology, including their involvement in politics, society, and culture.
Quarterly: Jan., Apr., July, Oct.
Yearly rates: Inst. $80 / Ind. $39 / ISSN: 0162-2439

SIMULATION & GAMING
An International Journal of Theory, Design, & Research
Official Journal of ABSEL, NASAGA, and ISAGA.
Editor: David Crookall, *Univ. of Alabama*
. . . publishes theoretical and empirical papers related to man, man-machine, and machine simulations of social processes; featured are theoretical papers about simulations in research and teaching, empirical studies, and technical papers on new gaming techniques.
Quarterly: March, June, Sept., Dec.
Yearly rates: Inst. $105 / Ind. $36 / ISSN: 1046-8781

SMALL GROUP RESEARCH **Name Chan**
An International Journal of Theory, Investigation, and Ap tion (Incorporating Small Group Behavior and Internat Journal of Small Group Research)
Editors: Charles Garvin, *Univ. of Michigan and*
Richard Brian Polley, *Lewis & Clark College*
. . . presents research, theoretical advancements, and empir supported applications with respect to all types of small gr Through advancing the systematic study of small groups, th terdiscipplinary journal seeks to increase communication ar all who are professionally interested in group phenomena
Quarterly: Feb., May, Aug. Nov.
Yearly rates: Inst. $98 / Ind. $38 / ISSN: 1046-4964

SMR/SOCIOLOGICAL METHODS AND RESEARCH
Editor: J. Scott Long, *Indiana Univ.*
. . . a leading journal of quantitative research and method in the social sciences.
Quarterly: Aug., Nov., Feb., May
Yearly rates: Inst. $100 / Ind. $38 / ISSN: 0049-1241

SOUTH ASIA JOURNAL
A Quarterly of the Indian Council for South Asian Cooper
Editor: Professor Bimal Prasad, *School of International St Jawaharlal Nehru Univ.*
. . . provides analyses of regional and national political, econ historical, and cultural issues among the nations of South
Quarterly: July, Oct., Jan., April
Yearly rates: Inst. $65 / Ind. $30 / ISSN: 0970-4868

STUDIES IN HISTORY
Editor: S. Gopal, *Centre for Historical Studies, Jawaharlal Nehru Univ., New Delhi*
. . . reflects the expansion and diversification that has occ in historical research in India in recent years.
Biannually: February and August
Yearly rates: Inst. $54 / Ind. $27 / ISSN: 0257-6430

URBAN AFFAIRS QUARTERLY
Editors: Dennis R. Judd and Donald Phares,
both at Univ. of Missouri, St. Louis
. . . emphasizes state-of-the-art research and scholarly an on urban themes: urban life, metropolitan systems, u economic development, and urban policy. Historical and c cultural perspectives add to its interdisciplinary features.
Quarterly: Sept., Dec., March, June
Yearly rates: Inst. $96 / Ind. $34 / ISSN: 0042-0816

URBAN EDUCATION
Editor: Warren Button, *SUNY Buffalo*
. . . exists to improve the quality of urban education by m the results of relevant empirical and scholarly inquiry from a ty of fields more widely available.
Quarterly: April, July, Oct., Jan.
Yearly rates: Inst. $98 / Ind. 34 / ISSN: 0042-0859

WESTERN JOURNAL OF NURSING RESEARCH
A Forum for Communicating Nursing Research
Editor: Pamela J. Brink, *Univ. of Alberta*
. . . an innovative forum for scholarly debate, as well a research and theoretical papers. Clinical studies have com taries and rebuttals. Departments deal with current issues in ing research.
Bimonthly: Feb., Apr., June., Aug., Oct., Dec.
Yearly rates: Inst. $108 / Ind. $48 ISSN: 0193-9459

WORK AND OCCUPATIONS
An International Sociological Journal
Editor: Curt Tausky, *Univ. of Massachusetts, Amherst*
. . . an international forum for sociological research and t in the substantive areas of work, occupations, leisure — structures and interrelationships.
Quarterly. Feb., May., Aug., Nov.
Yearly rates: Inst. $90 / Ind. $34 / ISSN: 0730-8884

WRITTEN COMMUNICATION
A Quarterly Journal of Research, Theory, & Applicati
Editors: Roger D. Cherry & Keith Walters, *Ohio State*
and Stephen P. Witte
. . . provides a forum for the free exchange of ideas, theo viewpoints, and methodological approaches that better defin further develop thought and practice in the exciting study written word.
Quarterly: Jan., April., July., Oct.
Yearly rates: Inst. $96 / Ind. $36 / ISSN: 0741-0883

YOUTH & SOCIETY
Editor: David Gottlieb, *Univ. of Houston*
. . . brings together interdisciplinary empirical studies theoretical papers on the broad social and political implic of youth culture and development; concentration is primar the age span from mid-adolescence through young adult
Quarterly: Sept., Dec., March., June
Yearly rates: Inst. $96 / Ind. $34 / ISSN: 0044-118X

JOURNAL OF PEACE RESEARCH
Published under the auspices of the International Peace Research Association
Editor: Nils Petter Gleditsch,
International Peace Research Institute Oslo
Quarterly: Feb., May, Aug., Nov.

JOURNAL OF SOCIAL AND PERSONAL RELATIONSHIPS
Editor: Steve Duck, *University of Iowa*
Quarterly: Feb., May, Aug., Nov.

JOURNAL OF THEORETICAL POLITI
Editors: Richard Kimber,
University of Keele,
Jan-Erik Lane, *University of Lund*
and Elinor Ostrom, *Indiana University*
Quarterly: Jan., April, July, Oct.

MEDIA, CULTURE & SOCIETY
Editors: Nicholas Garnham, Paddy Scannell, Colin Sparks,
Polytechnic of Central London,
Philip Schlesinger,
Thames Polytechnic
John Corner, *University of Liverpool*
and Nancy Wood, *University of Susse*
Quarterly: Jan., April, July, Oct.

SAGE RACE RELATIONS ABSTRACT
Published on behalf of the Institute of Race Relations, London
Editor: Louis Kushnick,
University of Manchester
Quarterly: Feb., May, Aug., Nov.

SCHOOL PSYCHOLOGY INTERNATIONAL
Editors: Robert L. Burden,
University of Exeter
and Caven S. Mcloughlin,
Kent State University, Ohio
Quarterly: Feb., May, Aug., Nov.

SOCIAL COMPASS
International Review of Sociology of Religion
Director: F. Houtard,
Editor: A. Bastenier,
University of Louvain, Belgium
Quarterly: March, June, Sept., Dec.

SOCIAL SCIENCE INFORMATION
Published under the auspices of the Maison des Sciences de l'Homme, with the collaboration of the Ecole des Hautes Etudes en Sciences Sociales, Paris
Editors: Elina Almasy and Anne Rocha-Perazzo
Quarterly: Feb., May, Aug., Nov.

SOCIAL STUDIES OF SCIENCE
An International Review of Research in the Social Dimensions of Science and Technology
Editor: David Edge,
University of Edinburgh,
Co-Editor: Roy MacLeod,
University of Sydney, Australia
Quarterly: Feb., May, Aug, Nov.

THEORY, CULTURE & SOCIETY
Explorations in Critical Social Scien
Editor: Mike Featherstone,
Teesside Polytechnic
Quarterly: Feb., May, Aug, Nov.

For more information and rates contact:

Sage Publications, Ltd.
Journal Marketing Dept.
28 Banner Street
London, EC1Y 8QE,
ENGLAND

U.S. and Canada inquiries to

Sage Publications, Inc.
PO Box 5096
Newbury Park, CA 91359

BUSINESS REPLY MAIL
FIRST CLASS MAIL PERMIT NO. 90 THOUSAND OAKS, CA

POSTAGE WILL BE PAID BY ADDRESSEE

SAGE PUBLICATIONS, INC.
P.O. Box 5084
NEWBURY PARK, CA 91359-9924